Vue.js 从入门到项目实战

胡同江　编著

清華大学出版社
北　京

内 容 简 介

本书从零基础开始讲解，用实例引导读者深入学习，深入浅出地讲解了Vue框架的各项实战技能。

本书共16章，主要讲解了Vue.js的基本概念，Vue实例和模板语法，计算属性、侦听器和过滤器，内置指令，页面元素样式的绑定，事件处理，双向数据绑定，组件技术，使用webpack打包，项目脚手架vue-cli，前端路由技术，状态管理等。最后列举了4个行业热点项目实训，包括神影视频App、音乐之家App、仿手机QQ页面、仿饿了么App。

本书适合任何想学习Vue框架的人员，无论您是否从事计算机相关行业，无论您是否接触过Vue框架，通过学习本书内容均可快速掌握Vue框架设计的方法和技巧。

图书在版编目(CIP)数据

Vue.js从入门到项目实战 / 胡同江编著. —北京：清华大学出版社，2019.10（2020. 12重印）
ISBN 978-7-302-53919-3

Ⅰ. ①V… Ⅱ. ①胡… Ⅲ. ①网页制作工具—程序设计 Ⅳ. ①TP393.092.2

中国版本图书馆 CIP 数据核字（2019）第 224350 号

责任编辑：张彦青
封面设计：李 坤
责任校对：李玉茹
责任印制：宋 林

出版发行：清华大学出版社
网 址：http://www.tup.com.cn，http://www.wqbook.com
地 址：北京清华大学学研大厦 A 座 **邮 编：**100084
社 总 机：010-62770175 **邮 购：**010-62786544
投稿与读者服务：010-62776969，c-service@tup.tsinghua.edu.cn
质 量 反 馈：010-62772015，zhiliang@tup.tsinghua.edu.cn
印 装 者：三河市少明印务有限公司
经 销：全国新华书店
开 本：185mm×260mm **印 张：**21.25 **字 数：**513 千字
版 次：2019 年 11 月第 1 版 **印 次：**2020 年 12 月第 3 次印刷
定 价：78.00 元

产品编号：082916-01

前　言

为什么要写这样一本书

Vue.js 是当下很火的一个 JavaScript MVVM 库，它是以数据驱动和组件化的思想构建的。Vue.js 提供了更加简洁、更易于理解的 API，能够在很大程度上降低 Web 前端开发的难度，因此深受广大 Web 前端开发人员的喜爱。Vue.js 库功能虽然强大，但是对于初学者来说入门比较困难，因此我们选择了让读者从入门到掌握 Vue.js 作为编写本书的思路，没有基础的读者通过本书学习也没有障碍。本书将项目开发中的技能融入本书的案例中，通过本书的学习，不仅可以掌握 Vue.js 库的使用方法，还可以积累项目开发经验，从而满足企业实际开发的需求。

本书特色

零基础、入门级的讲解

无论您是否从事计算机相关行业，无论您是否接触过 Vue 框架，都能从本书中找到最佳起点。

实用、专业的范例和项目

本书在编排上紧密结合深入学习 Vue 框架技术的过程，从 Vue 框架基本操作开始，逐步带领读者学习 Vue 框架的各种应用技巧，侧重实战技能，使用简单易懂的实际案例进行分析和操作指导，让读者学起来简明轻松，操作起来有章可循。

随时随地学习

本书提供了微课视频，通过手机扫码即可观看，随时随地解决学习中的困惑。

细致入微、贴心提示

本书在讲解过程中，在各章中使用了“注意”“提示”“技巧”等小栏目，使读者在学习过程中能更清楚地了解相关操作、理解相关概念，并轻松掌握各种操作技巧。

超值资源大放送

全程同步教学录像

涵盖本书所有知识点，详细讲解每个实例及项目的过程及技术关键点。比看书更能轻松地掌握书中所有的网页制作和设计知识，而且扩展的讲解部分可使您得到比书中更多的收获。

超多容量王牌资源

赠送大量王牌资源，包括实例源代码、教学幻灯片、本书精品教学视频、实用网页模板、网页开发必备参考手册、HTML 5 标签速查手册、精选的 JavaScript 实例、CSS 3 属性速查表、JavaScript 函数速查手册、CSS+DIV 布局案例赏析、精彩网站配色方案赏析、网页样式与布局案例赏析、Web 前端工程师常见面试题等。读者可以在清华大学出版社官网搜索本书获取赠送资源。

读者对象

- 没有任何 Vue 框架开发基础的初学者。
- 有一定的 Vue 框架开发基础，想精通前端框架开发的人员。
- 有一定的网页前端设计基础，没有项目经验的人员。
- 大专院校及培训学校的老师和学生。

创作团队

本书由胡同江编著，参加编写的人员还有刘春茂、李艳恩和李佳康。在编写过程中，我们虽竭尽所能将最好的讲解呈现给了读者，但难免有疏漏和不妥之处，敬请读者不吝指正。

编 者

目 录

Contents

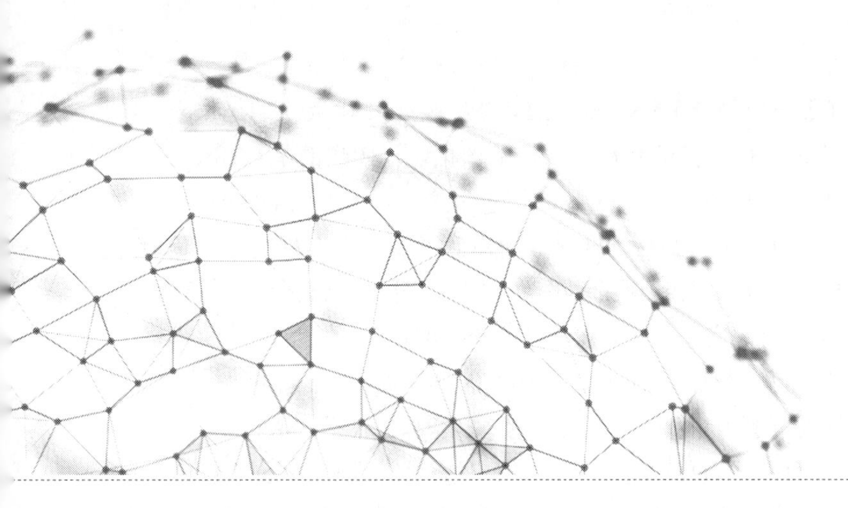

第1章 Vue.js简介与安装

很多人都说 Web 前端开发是程序员中门槛最低的，5 天就可以从入门到精通，你是否有“道理我都懂，就是不知道怎么入手”的困惑？你真的知道如何做好一个前端开发者吗？其实前端开发远不止切图、写 HTML、调样式，前端的世界很大，Node.js、webpack 和各种各样的前端框架都是 Web 前端世界的一部分。

本章主要介绍与 Vue.js（以下简称 Vue）有关的一些概念与技术，并帮助读者了解它们背后相关的工作原理。通过对本章的学习，即使从未接触过 Vue，也可以运用这些知识点快速构建出一个 Vue 应用。

1.1 Vue 概述

说到前端框架，当下比较流行的有3个：Vue、React.js（以下简称React）和Angular.js（以下简称 Angular）。其中 Vue 以其容易上手的 API、不俗的性能、渐进式的特性和活跃的社区，在三大框架中脱颖而出，截至 2019 年 3 月，Vue 在 github 上的 star 数已经超过了其他两个框架，成为三大框架中最热门的框架。

Vue 被定义成一个用来开发 Web 界面的前端框架，是个轻量级的工具。使用 Vue 可以让 Web 开发变得简单，同时也颠覆了传统前端开发模式。它提供了现代 Web 开发中常见的高级功能，例如：

- 逻辑视图与数据。
- 可复用的组件。
- 前端路由。
- 状态管理。
- 虚拟 DOM（Virtual DOM）。

1.1.1 MVVM 模式

学习 Vue 之前，先来了解一下 MVVM 模式。MVVM 是 Model-View-ViewModel 的简写，即模型 - 视图 - 视图模型。模型指的是后端传递的数据，视图指的是 HTML 页面，视图模型是 MVVM 模式的核心，它是连接 View 和 Model 的桥梁。MVVM 有两个方向：一是将模型转化成视图，即将后端传递的数据转化成所看到的页面，实现的方式是数据绑定。二是将视图转化成模型，即将所看到的页面转化成后端的数据，实现的方式是 DOM 事件监听。如果这两个方向都实现，称之为数据的双向绑定。

在 MVVM 的框架中，视图和模型是不能直接通信的，它们通过 ViewModel 来通信，ViewModel 通常要扮演一个监听者的角色，当数据发生变化时，ViewModel 能够监听到数据的变化，然后通知对应的视图做自动更新；而当用户操作视图时，ViewModel 也能监听到视图的变化，然后通知数据做改动，这实际上就实现了数据的双向绑定。并且 MVVM 中的 View 和 ViewModel 可以互相通信。MVVM 流程图如图 1-1 所示。

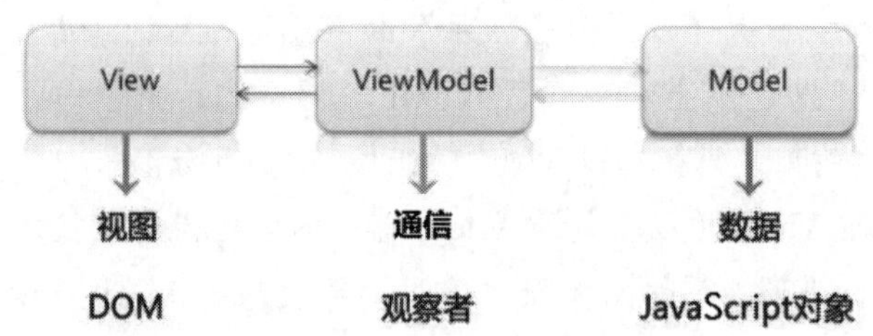

图 1-1　MVVM 流程图

Vue 就是基于 MVVM 模式实现的一套框架。在 Vue 中，Model 指的是 JavaScript 中的数据，例如对象、数组等，View 指的是页面视图，ViewModel 指的是 Vue 实例化对象。

1.1.2 Vue 是什么

在过去的十年里，我们的网页变得更加动态化和强大了，因为有 JavaScript，我们可以把很多传统的服务端代码放到浏览器中，这样就产生了成千上万行的 JavaScript 代码，它们连接了各式各样的 HTML 和 CSS 文件，但缺乏正规的组织形式，这也是为什么越来越多的开发者使用 JavaScript 框架的原因，例如 Vue、Angular 和 React。

Vue 是一款友好的、多用途且高性能的 JavaScript 框架，它能够帮助你创建可维护性和可测试性更强的代码库。Vue 是一个渐进式的 JavaScript 框架，渐进式的意义如下：

（1）用户可以一步一步、有阶段性地来使用 Vue，不必一开始就使用所有的东西。

（2）如果已经有一个现成的服务端应用，可以将 Vue 作为该应用的一部分嵌入其中，带来更加丰富的交互体验。

（3）如果希望将更多业务逻辑放到前端来实现，那么 Vue 的核心库及其生态系统也可以满足用户的各种需求。和其他前端框架一样，Vue 允许用户将一个网页分割成可复用的组件，每个组件都包含属于自己的 HTML、CSS、JavaScript 以用来渲染网页中相应的地方。

（4）如果用户构建一个大型的应用，可能需要将东西分割成为各自的组件和文件，Vue 有一个命令行工具，使快速初始化一个真实的工程变得非常简单。

可以看出，Vue 的使用可大可小，它都会有相应的方式来整合到用户的项目中。所以说它是一个渐进式的框架。

Vue 本身具有响应式编程和组件化的特点。

响应式：即为保持状态和视图的同步，也被称为数据绑定，声明实例 new Vue({data:data}) 后自动对 data 里面的数据进行了视图上的绑定，修改 data 的数据，视图中对应数据也会随之更改。

组件化：Vue 组件化的理念和 React 异曲同工——“一切都是组件”。可以将任意封装好的代码注册成标签，例如：Vue.component('example',Example)，可以在模板中以 <example></example> 的形式调用。如果组件设计合理，在很大程度上能减少重复开发，而且配合 Vue 的插件 vue-loader，可以将一个组件的 CSS、HTML 和 JavaScript 都写在一个文件里，做到模块化的开发。除此之外，Vue 也可以与 vue-router 和 vue-resource 插件配合起来，以支持路由和异步请求，这样就满足了开发单页面应用的基本条件。

提示

Vue 不支持 IE 8 及以下版本，因为 Vue 使用了 IE 8 无法模拟的 ECMAScript 5 特性，但它支持所有兼容 ECMAScript 5 的浏览器。

1.1.3 Vue 有什么不同

在前端开发中，会遇到动画、交互效果、页面特效等业务，原生的 JavaScript 或 jQuery 库通过操作 DOM 来实现，数据和界面是连接在一起的，例如下面的示例。

在示例中添加一段文本和一个按钮，并为按钮添加一个单击事件，当单击按钮时把文本中的“王老师”更改为“李老师”，“30”更改为“40”。

```
<div>
    <p>大家好,我是<span id="name">王老师,</span></p>
    <p>今年<span id="age">30</span>岁。</p>
    <button id = "updata">修改</button>
</div>
<script>
    $("#updata").click(function(){
        $("#name").text("李老师");
        $("#age").text("40");
    });
</script>
```

在 IE 11 浏览器中运行，效果如图 1-2 所示，单击“修改”按钮时，页面中的内容更改，结果如图 1-3 所示。

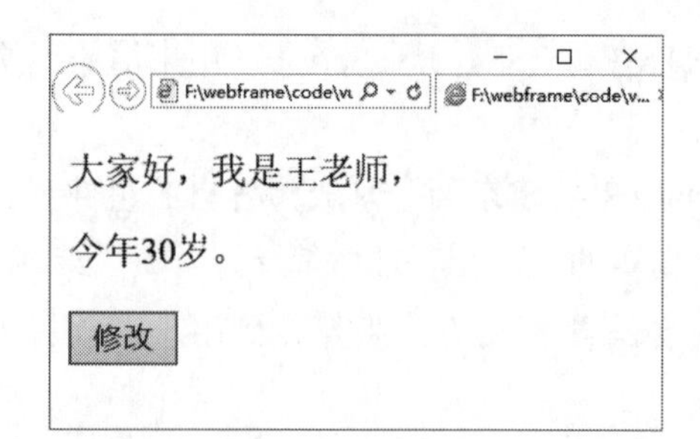

图 1-2　初始化效果

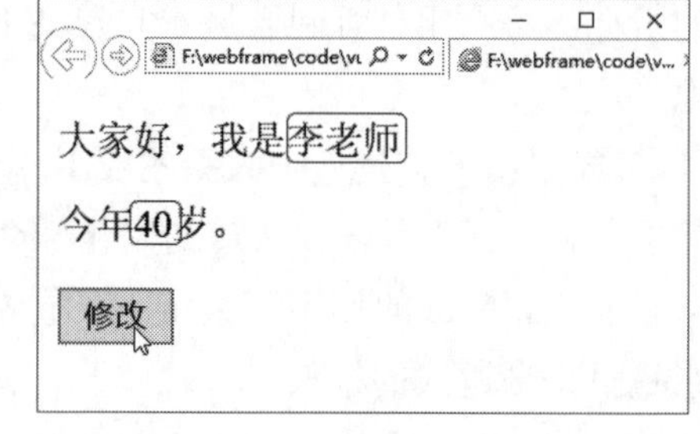

图 1-3　单击“修改”按钮后的效果

Vue 将数据层和视图层完全分离开，不仅对 API 进行封装，还提供了一系列的解决方案。这是一个思想的转变，数据驱动的机制，主要操作的是数据而不是频繁地操作 DOM（导致页面频繁重绘）。使用 Vue 实现上述示例，代码如下：

```
<div id="app">
    <p>大家好,我是<span>{{name}}</span></p>
    <p>今年<span>{{age}}</span>岁。</p>
    <button v-on:click="updata">修改</button>
</div>
<script>
    new Vue({
        el: '#app',
        data:{
            name:"王老师",
            age:"30"
        },
        methods:{
            updata:function(){
                this.name = "李老师";
                this.age = "40";
            }
        }
    })
</script>
```

提示

对于上面的 Vue 代码，暂时不用理解，这里只是快速展示 Vue 的写法，详细的介绍请参考下面“1.4 第一个 Vue 程序”。

以上示例总结如下。

（1）jQuery 首先要获取到 DOM 对象，然后对 DOM 对象进行值的修改等操作。

（2）Vue 首先把值和 JavaScript 对象进行绑定，然后修改 JavaScript 对象的值，Vue 框架就会自动把 DOM 的值进行更新。

（3）可以简单地理解为 Vue 帮我们做了 DOM 操作，以后使用 Vue 只需要修改对象的值以及做好元素和对象的绑定，Vue 框架就会自动帮我们做好 DOM 的相关操作。

（4）如果 DOM 元素跟随 JavaScript 对象值的变化而变化，叫作单向数据绑定；如果 JavaScript 对象的值也跟随着 DOM 元素的值的变化而变化，叫作双向数据绑定。

1.2 如何使用 Vue.js

Vue 是当下流行的前端框架之一，在正式开始学习它前，我们先对传统前端开发模式和 Vue 的开发模式做一个对比，以此了解 Vue 产生的背景和 Vue 核心思想。

1.2.1 传统的前端开发模式

传统开发模式可称为硬代码开发，数据、展现和逻辑都混在一起，彼此相互混杂，整体看起来非常混乱，它有以下缺点。

（1）由于展现、数据和逻辑都混合到一起，从而代码的可读性就会很差，很难完成知识的转移和交付。

（2）界面变更修改复杂，无法快速调试，问题出来了，无法快速定位问题所在。

（3）维护复杂，容易在修复中出现新 bug。

（4）数据处理功能单一，若出现排序、筛选等工作，需要重新编写代码。

前端技术在近几年发展迅速，如今的前端开发已不再是 10 年前写个 HTML 和 CSS 那样简单了，新的概念层出不穷，例如 ECMAScript 6、Node.js、NPM、前端工程化等。这些新东西在不断优化我们的开发模式，改变我们的编程思想。

随着项目的扩大和时间的推移，出现了更复杂的业务场景，例如组件解耦、SPA（单页面应用）等。为了提升开发效率，降低维护成本，传统的前端开发模式已不能完全满足我们的需求，这时就出现了 Angular、React 以及我们要介绍的 Vue。

1.2.2 Vue.js 的开发模式

Vue 是基于 MVVM 模式实现的一套框架，MVVM 模式分离视图（View）和数据（Model），通过自动化脚本实现自动化关联，ViewModel 搭起了视图与数据的桥梁，同时在 ViewModel 里进行交互及逻辑处理。可以简单地理解为，View 就是 HTML、DOM，数据 Model 就是要处理的 Json 数据。这种模式有以下优势。

（1）低耦合。将 View 和 Model 进行分离，两者中其中一方变更时，另一方不会受到影响。

（2）重用性。无论是 View、ViewModel 还是 Model，三者都可以进行重用，提高了开发效率。

（3）HTML 模板化。修改模板不影响逻辑和数据，模板可直接调试。

（4）数据自动处理。Model 实现了标准的数据处理封装，例如排序、筛选等。

（5）双向绑定。通过 DOM 和 Model 双向绑定使数据更新自动化，缩短了开发时间。

1.3 安装 Vue

在学习 Vue 之前，先来学习一下如何安装它。

1.3.1 直接用 <script> 引入

直接使用 <script> 标签引入有两种方式，一种是官网下载独立的版本，另一种是使用 CDN 的方式。本书前面内容都是使用独立的版本来进行介绍，推荐读者也用这种方式；而本书后面的项目，是使用 Vue 的脚手架进行创建。

1. 独立的版本

在官网 https://cn.vuejs.org/ 下载最新稳定版本：2.6.10，使用 <script> 标签引入到页面中，这时 Vue 会被注册为一个全局变量。

官网提供了两个版本，说明如图 1-4 所示。根据需要下载相应的版本，本书使用的是开发版本 2.6.10。

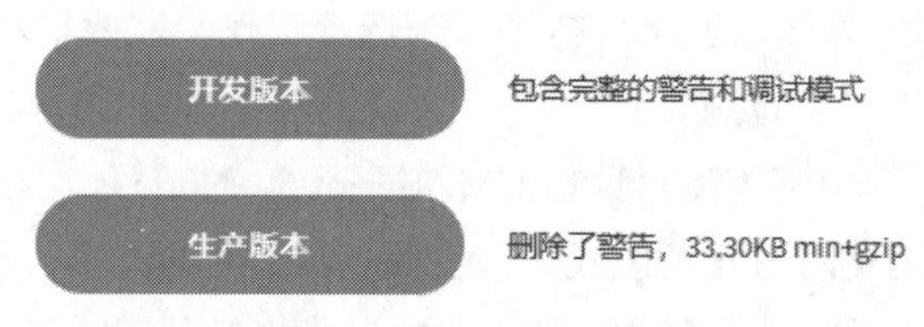

图 1-4　Vue 的下载版本

注意

在开发环境下不要使用生产版本，不然就失去了所有常见错误相关的警告。

下载完成后直接引用：

```
<script src="vue.js"></script>
```

2. CDN 方式

对于制作原型或学习，可以使用最新版本：

```
<script src="https://cdn.jsdelivr.net/npm/vue"></script>
```

对于生产环境，推荐链接到一个明确的版本号和构建文件，以避免新版本产生的不可预期的问题，例如：

```
<script src="https://cdn.jsdelivr.net/npm/vue@2.6.10/dist/vue.js"></script>
```

1.3.2 NPM

在用 Vue 构建大型应用时推荐使用 NPM 安装。NPM 能很好地和诸如 webpack 或 Browserify 模块打包器配合使用。同时 Vue 也提供了配套工具来开发单文件组件。

由于 NPM 安装速度慢，推荐使用淘宝 NPM 镜像 CNPM。

```
# 最新稳定版
$ npm install vue
```

提示

对于中国用户，建议将 NPM 源设置为国内的淘宝 NPM 镜像，可以大幅提升安装速度。

1.3.3 命令行工具

Vue 提供了一个官方的脚手架（CLI），为单页面应用（SPA）快速搭建繁杂的脚手架。它为现代前端工作流提供了 batteries-included 的构建设置，只需要几分钟的时间就可以运行起来并带有热重载、保存时 lint 校验以及生产环境可用的构建版本。

CLI 工具假定用户对 Node.js 和相关构建工具有一定程度的了解。如果是新手，建议先在熟悉 Vue 本身之后再使用 CLI。本书后面章节将具体介绍脚手架的安装以及如何快速创建一个项目。

1.4 第一个 Vue 程序

引入 Vue 框架后，在 <body> 底部使用 new Vue() 的方式创建一个实例，就可以开始使用 Vue 了。下面通过一个完整的示例来看一下 Vue 的实现过程。

【例 1.1】第一个 Vue 程序。

```
<!DOCTYPE html>
<html>
<head>
    <meta charset="UTF-8">
    <title></title>
    <!--引入Vue框架-->
    <script src="https://cdn.jsdelivr.net/npm/vue@2.6.10/dist/vue.js"></script>
</head>
<body>
<div id="app">
    <p>大家好,我是<b>{{name}},</b></p>
    <p>今年<b>{{age}}</b>岁。</p>
</div>
<script>
```

```
    //创建实例
    var app=new Vue({
        el: '#app',
        data:{
            name:"王老师",
            age:"30"
        }
    })
</script>
</body>
</html>
```

在 IE 11 浏览器里面运行的结果如图 1-5 所示。

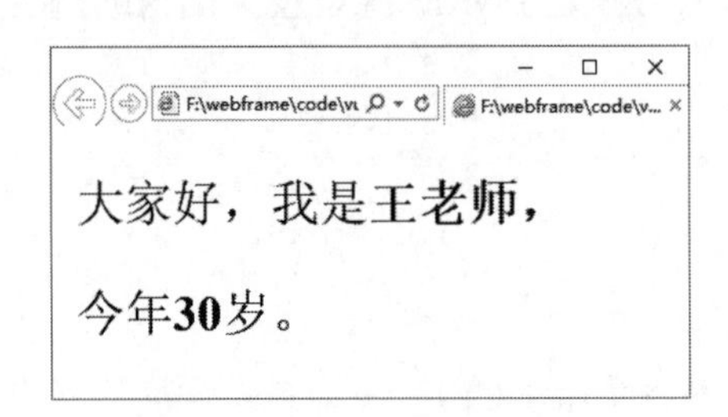

图 1-5　第一个 Vue 程序效果

在后面的章节中还会学习到一些其他选项，例如计算属性 computed、方法 methods 等。

提示

在创建的 Vue 实例中，el 选项用于指定一个页面中已经存在的 DOM 元素来挂载 Vue 实例，它可以是 HTMLElement，也可以是 CSS 选择器。data 选项用于声明应用内需要双向绑定的数据。建议所有会用到的数据都预先在 data 中声明，这样不至于将数据散落在业务逻辑中，难以维护。

Vue 实例代理了 data 对象里的所有属性，所以可以像上面例子那样进行访问。除了显式的生命数据外，也可以指向一个已有的变量，并且它们之间默认建立了双向绑定。

```
<script>
    var myData ={
        a: 123
    }
    //创建实例
    var app = new Vue({
        el: '#app',
        data: myData
    })
    console.log(app.a);
    app.a = 456;
    console.log(myData.a);
</script>
```

在 IE 11 浏览器中运行，并打开控制台，app.a 的打印结果为 123；当更改 app.a 的值后，原数据也发生改变，为 456，结果如图 1-6 所示。

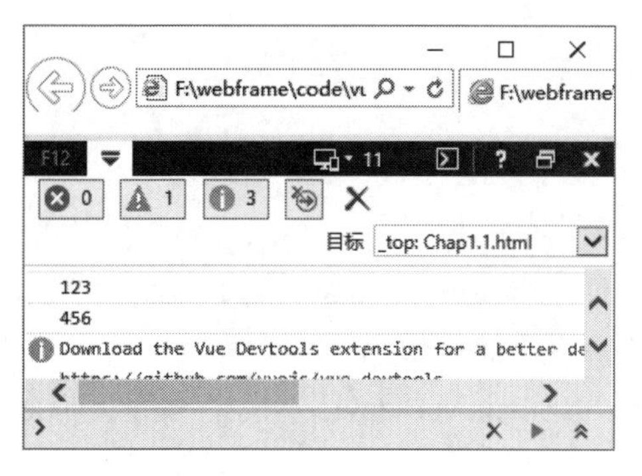

图 1-6　控制台打印结果

以上就成功创建了第一个 Vue 应用，看起来这跟渲染一个字符串模板非常类似，但是 Vue 在背后做了大量工作。现在数据和 DOM 已经被建立了关联，所有东西都是响应式的。

我们可以通过浏览器的 JavaScript 控制台来验证（就在这个页面打开），例如修改 app.name=" 李老师 "、app.age="40"，如图 1-7 所示；当单击“执行”按钮后，将看到上例相应地更新，如图 1-8 所示。

图 1-7　修改代码

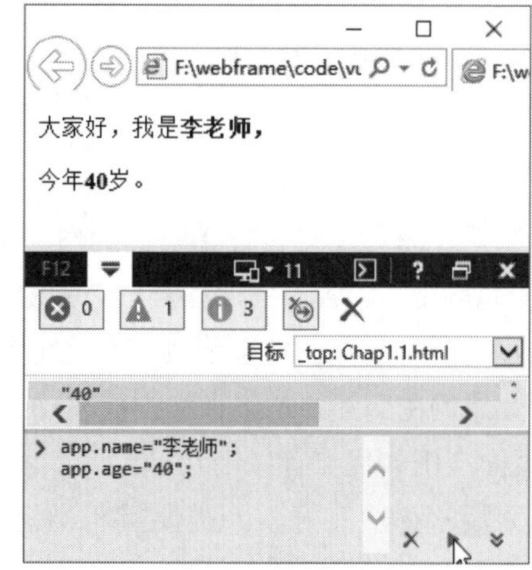

图 1-8　执行后效果

注意

app.name=" 李老师 " 和 app.age="40" 中的 app 是我们创建的实例名称。

在以后的章节中，示例不再提供完整的代码，而是根据上下文，将 HTML 部分与 JavaScript 部分单独展示，省略了 <head>、<body> 等标签以及 Vue 的加载等内容，读者可根据上例结构来组织代码。

1.5 疑难解惑

疑问 1：Vue Devtools 工具是什么？

Vue Devtools 是基于 Google Chrome 浏览器的一款调试 Vue 应用的开发者浏览器扩展，可以在浏览器开发者工具下调试代码。

可访问 https://github.com/vuejs/vue-devtools 进行安装，界面效果如图 1-9 所示。

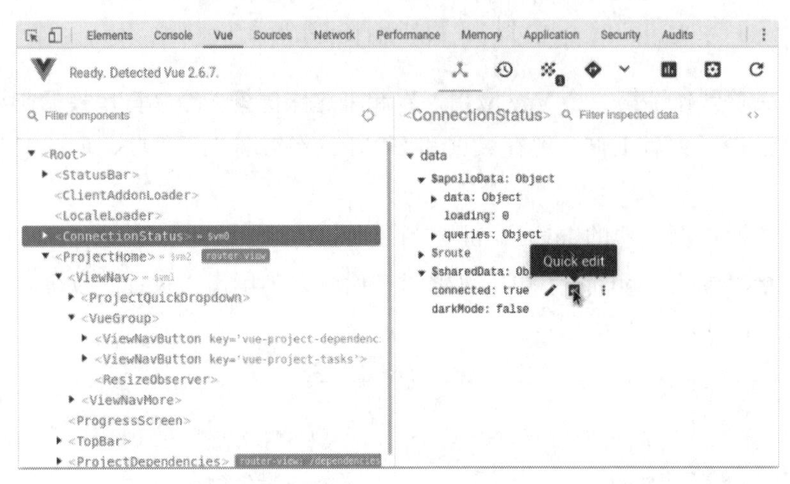

图 1-9　界面效果

疑问 2：Vue 很适合创建单页面应用，那什么是单页面应用？优缺点是什么？

单页面应用（SPA）是指只有一个主页面的应用，浏览器一开始要加载所有必需的 html、js、css。所有的页面内容都包含在这个主页面中。但在编写的时候，还是会分开写（页面片段），然后在交互的时候由路由程序动态载入，单页面的页面跳转，仅刷新局部资源。

单页面应用的优点有以下几个。

（1）用户体验良好。不需要重新刷新页面，减少 TTFB 的请求耗时，数据也是通过 Ajax 异步获取，页面显示流畅。

（2）前后端分离。前端负责界面显示，后端负责数据存储和计算，各司其职，不会把前后端的逻辑混杂在一起。

（3）减轻服务端压力，服务器只需要提供 API 接口，不用管理页面逻辑和页面的拼接，吞吐能力会提高几倍。

（4）共用一套后端程序代码，适配多端同一套后端程序代码，不用修改就可以适用于 Web、手机、平板。

单页面应用的缺点有以下几个。

（1）不利于 SEO 的优化。因为页面数据都是前端异步加载的方式，不利于搜索引擎的抓取。

（2）首屏加载过慢。单页面首次加载，需要将所有页面所依赖的 CSS 和 JS 合并后统一加载，所以 CSS 和 JS 文件会较大，影响页面首次打开的时间。

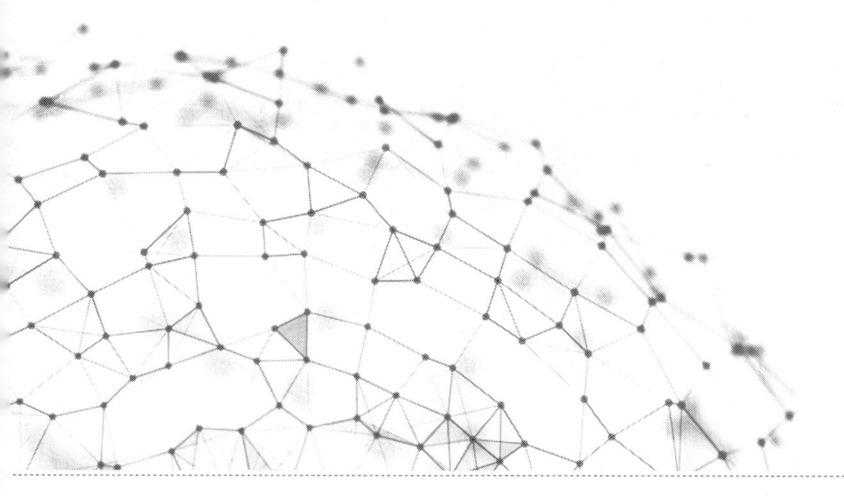

第2章

Vue实例和模板语法

在第 1 章中，已经介绍了 Vue 的基础知识以及第一个 Vue 程序，本章将介绍如何创建 Vue 实例和模板语法，这是新手入门必须要掌握的知识点。

2.1 Vue 实例

无论是用官方的脚手架，还是自己搭建的项目模板，最终都会创建一个 Vue 的实例对象并挂载到指定 DOM 上。下面来介绍一下关于 Vue 实例的相关内容。

2.1.1 创建一个 Vue 实例

每个 Vue 应用都是通过用 Vue 函数创建一个新的 Vue 实例开始的：

```
var app = new Vue({
  //选项
})
```

当创建一个 Vue 实例时，可以传入一个选项对象，这些选项用来创建想要的行为（methods、computed、watch 等）。

一个 Vue 应用由一个通过 new Vue 创建的根 Vue 实例以及可选的、嵌套的、可复用的组件树组成。例如，一个 Todo 应用的组件树如图 2-1 所示。

```
根实例
└ TodoList
   ├ TodoItem
   │ ├ DeleteTodoButton
   │ └ EditTodoButton
   └ TodoListFooter
      ├ ClearTodosButton
      └ TodoListStatistics
```

图 2-1 组件树

在后面的组件系统章节会具体展开介绍。读者现在只需要明白所有的 Vue 组件都是 Vue 实例，并且接受相同的选项对象（一些根实例特有的选项除外）。

2.1.2 数据与方法

当一个 Vue 实例被创建时，它将 data 对象中的所有属性加入到 Vue 的响应式系统中。当这些属性的值发生改变时，视图将会产生“响应”，即匹配更新为新的值。

【例 2.1】数据响应。

```
<div id="app"></div>
<script>
    //我们的数据对象
    var data = { a: 1 }
    var app=new Vue({
        data: data
    })
    //获得Vue实例上的属性，返回源数据中对应的字段
    app.a == data.a  //=>true
    //设置属性也会影响到原始数据
    app.a = 2
    data.a  // => 2
    console.log(data.a)   //打印此刻原始数据的值
    //反之亦然
    data.a = 3
    app.a // => 3
    console.log(app.a)   //打印此刻原始数据的值
</script>
```

在 IE 11 浏览器里面运行的结果如图 2-2 所示。

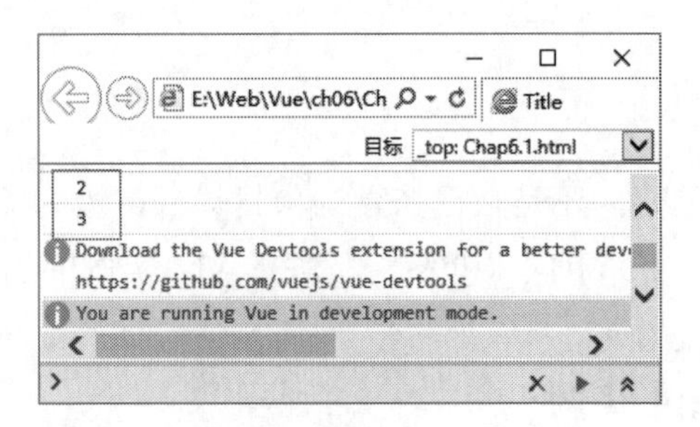

图 2-2 数据响应

当这些数据改变时，视图会进行重新渲染。要注意的是，只有当实例被创建时，data 中存在的属性才是响应式的。也就是说，如果添加一个新的属性，例如：

```
app.b=10;
```

那么对 b 的改动将不会触发任何视图的更新。如果后期需要一个属性，但是一开始它为空或不存在，可以先定义它，只需要设置一些初始值即可。例如：

```
data: {
  newTodoText: '',
  visitCount: 0,
  hideCompletedTodos: false,
  todos: [],
  error: null
}
```

唯一的例外是使用 Object.freeze() 方法，该方法会阻止修改现有的属性，也意味着响应系统无法再追踪变化。

【例 2.2】 Object.freeze() 方法。

```
<div id="app">
    <p>{{ foo }}</p>
    <!-- 这里的 foo 不会更新! -->
    <button v-on:click="foo = '100'">改变它</button>
</div>
<script>
    var obj = {
        foo: '10'
    }
    Object.freeze(obj);  //冻结obj对象
    new Vue({
        el: '#app',
        data: obj
    })
</script>
```

在 IE 11 浏览器里面运行，然后单击“改变它”按钮，可以发现 foo 的值并不会改变，效果如图 2-3 所示。

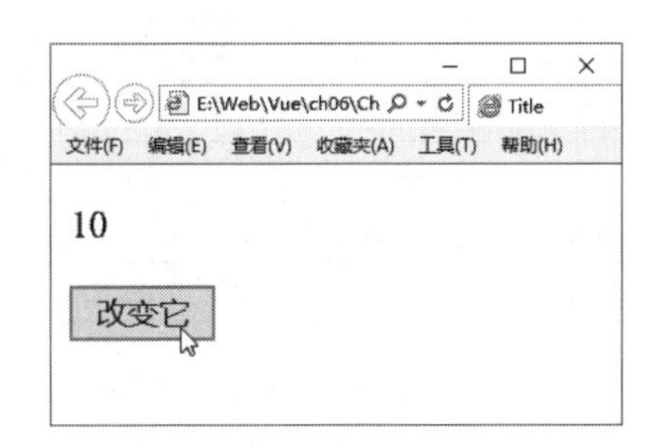

图 2-3 Object.freeze() 方法的作用效果

其中“v-on:click=”用来定义单击事件，在后面的章节中会具体进行介绍。

提示

Object.freeze() 方法可以冻结一个对象，一个被冻结的对象再也不能被修改。冻结了一个对象后不能向这个对象添加新的属性，不能删除已有属性，不能修改该对象已有属性的可枚举性、可配置性、可写性，也不能修改已有属性的值。此外，冻结一个对象后该对象的原型也不能被修改。

除了数据属性外，Vue 实例还有一些有用的实例属性与方法。它们都有前缀 $，以便与读者自定义的属性区分开来。例如：

```
<script>
    var data = { a: 1 }
    var app= new Vue({
        el: '#example',
        data: data
```

```
    })
    app.$data === data // => true
    app.$el === document.getElementById('example') // => true
    //$watch 是一个实例方法
    app.$watch('a', function (newValue, oldValue) {
        //这个回调函数将在 app.a 改变后调用
    })
</script>
```

2.1.3 实例生命周期钩子

每个Vue实例在被创建时，都要经过一系列的初始化过程。例如，需要设置数据监听、编译模板、将实例挂载到DOM，并在数据变化时更新DOM等。同时在这个过程中也会运行一些叫作生命周期钩子的函数，这给了开发者在不同阶段添加自己的代码的机会。

常用生命周期钩子函数如表2-1所示。

表2-1 钩子函数

钩子函数	说　明
beforeCreate	在Vue实例创建之前执行的函数
created	实例创建完成后调用
beforeMount	在Vue实例创建之后，数据未渲染时负责接管DOM之前执行的函数
mounted	el挂载到实例上后调用，一般我们的第一个业务逻辑会在这里开始
beforeDestory	实例销毁之前调用。主要解绑一些使用addEventListener监听的事件
destroyed	Vue实例在执行vm.destroyed()命令之后，销毁之后执行的函数
beforeUpdate	在Vue实例数据更新之前执行的函数
updated	在Vue实例数据更新之后执行的函数

这些钩子与el和data类似，也是作为选项写入Vue实例内，并且钩子的this指向的是调用它的Vue实例。

2.1.4 实例化多个对象

实例化多个Vue对象和实例化单个Vue对象方法一样，只是绑定操控的el元素不同了。例如，创建两个Vue对象，分别命名为one和two，实例化这两个对象的案例如下。

【例2.3】实例化两个Vue对象。

```
<h3>初始化多个Vue实例对象</h3>
<div id="app-one">
    <h4>{{title}}</h4>
</div>
<div id="app-two">
    <h4>{{title}}</h4>
</div>
<script>
    //实例化对象一
    var one=new Vue({
        el: '#app-one',
```

```
        data:{
            title:'app-one的内容'
        }
    });
    //实例化对象二
    var two=new Vue({
        el: '#app-two',
        data:{
            title:'app-two的内容'
        }
    })
</script>
```

在 IE 11 浏览器里面运行，可看到两个 Vue 对象 data 属性的内容，效果如图 2-4 所示。

图 2-4　两个 Vue 对象 data 属性的内容

除了可以展示 data 属性的内容，也可以展示 computed 计算属性的内容。

【例 2.4】展示 computed 计算属性的内容。

```
<h1>初始化多个Vue实例对象</h1>
<div id="app-one">
    <h2>{{title}}</h2>
    <p>{{say}}</p>
</div>
<div id="app-two">
    <h2>{{title}}</h2>
    <p>{{say}}</p>
</div>
<script>
    //实例化对象一
    var one=new Vue({
        el: '#app-one',
        data:{
            title:'app-one的内容'
        },
        computed:{
            say:function(){
                return 'Hello One';
            }
        }
    });
    //实例化对象二
    var two=new Vue({
        el: '#app-two',
        data:{
```

```
            title:'app-two的内容'
        },
        computed:{
            say:function(){
                return 'Hello Two';
            }
        }
    })
</script>
```

在 IE 11 浏览器里面运行，可看到两个 Vue 对象 computed 计算属性的内容，效果如图 2-5 所示。

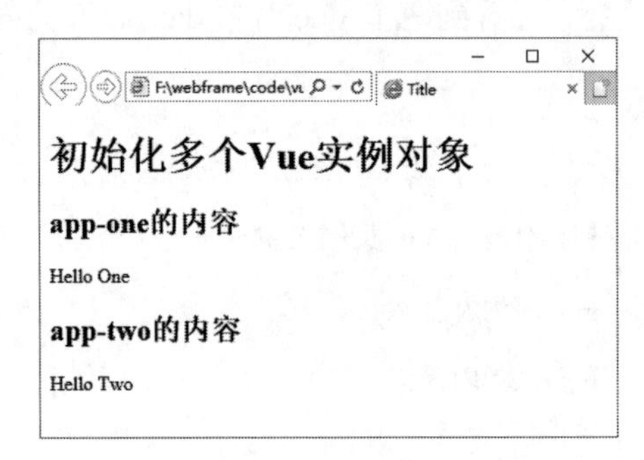

图 2-5　两个 Vue 对象 computed 计算属性的内容

如果想在第二个实例化对象中改变第一个实例化对象中的 data 属性，可以在第二个实例化对象中定义一个方法，通过事件来触发，在该事件中调用第一个对象，更改其中的 title 属性。

例如，在上面示例的第二个对象中，定义一个方法 changeTitle，在方法中调用第一个对象，并修改其中的 title 属性。

【例 2.5】改变其他实例中的属性内容。

```
<div id="app-one">
    <h2>{{title}}</h2>
    <p>{{say}}</p>
</div>
<div id="app-two">
    <h2>{{title}}</h2>
    <p>{{say}}</p>
    <button v-on:click="changeTitle">改变第一个对象中title属性</button>
</div>
<script>
    //实例化对象一
    var one=new Vue({
        el: '#app-one',
        data:{
            title:'app-one的内容'
        },
        computed:{
            say:function(){
                return 'Hello One';
            }
        }
```

```
    });
    //实例化对象二
    var two=new Vue({
        el: '#app-two',
        data:{
            title:'app-two的内容'
        },
        methods:{
            changeTitle:function(){
                one.title='已经改名了！';
            }
        },
        computed:{
            say:function(){
                return 'Hello Two';
            }
        }
    })
</script>
```

在 IE 11 浏览器里面运行，效果如图 2-6 所示。当单击“改变第一个对象中 title 属性”按钮后，实例对象 one 中的 title 内容将发生改变，如图 2-7 所示。

图 2-6　页面初始化效果

图 2-7　实例 one 改变属性后的效果

还可以在实例化对象外面调用，去更改它的属性，例如改变实例 two 的 title 属性：

```
two.title='实例化对象二的内容已经改变了';
```

在 IE 11 浏览器中显示效果如图 2-8 所示。

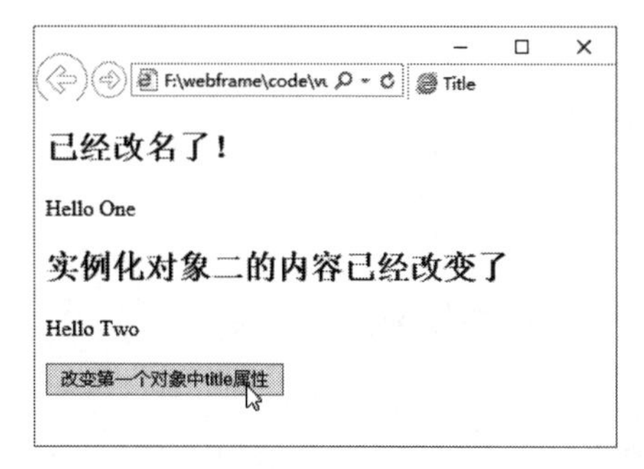

图 2-8　实例 two 改变属性后的效果

2.2 模板语法

Vue 使用了基于 HTML 的模板语法，允许开发者声明式地将 DOM 绑定至底层 Vue 实例的数据上。所有 Vue 的模板都是合法的 HTML，所以能被遵循规范的浏览器和 HTML 解析器解析。

在底层的实现上，Vue 将模板编译成虚拟 DOM 渲染函数。结合响应系统，Vue 能够智能地计算出最少需要重新渲染多少组件，并把 DOM 操作次数减到最少。

2.2.1 插值

插值的语法有以下 3 种。

1. 文本

数据绑定最常见的形式就是使用“Mustache”语法（双大括号）的文本插值：

```
<span>Message: {{ message}}</span>
```

Mustache 标签将会被替代为对应数据对象上 message 属性的值。无论何时，绑定的数据对象上 message 属性发生了改变，插值处的内容都会更新。

通过使用 v-once 指令，也能执行一次性地插值，当数据改变时，插值处的内容不会更新。但这会影响到该节点上的其他数据绑定：

```
<span v-once>这个将不会改变: {{ message }}</span>
```

在下面示例中，在标题中插值，插值为“Vue.js”，可以根据需要进行修改。

【例 2.6】渲染文本。

```
<div id="app">
    <h3>本书教大家如何学习{{message}}</h3>
</div>
<script>
    new Vue({
        el: '#app',
        data:{
            message:'Vue.js'
        }
    })
</script>
```

在 IE 11 浏览器里面运行的结果如图 2-9 所示。

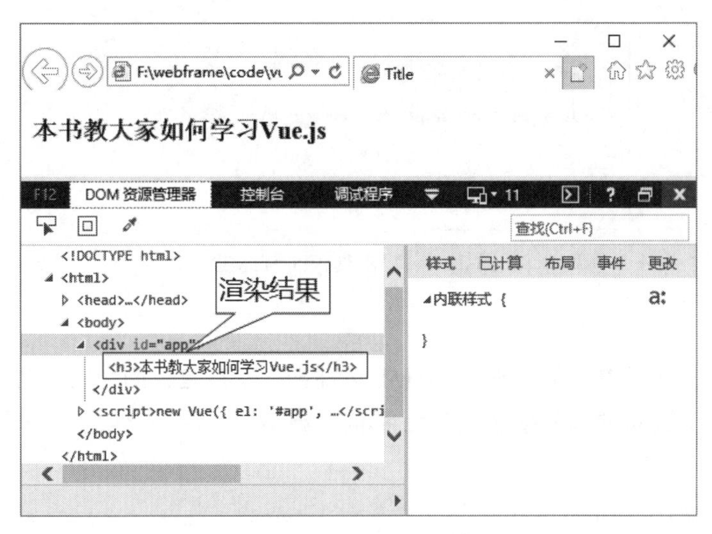

图 2-9 文本渲染效果

2. 原始 HTML

双大括号会将数据解释为普通文本，而非 HTML 代码。为了输出真正的 HTML，我们需要使用 v-html 指令。

提示

不能使用 v-html 来复合局部模板，因为 Vue 不是基于字符串的模板引擎。对于用户界面（UI），组件更适合作为可重用和可组合的基本单位。

例如，想要输出一个 a 标签，首先需要在 data 属性中定义该标签，再根据需要定义 href 属性值和标签内容。然后使用 v-html 绑定到对应的元素上。

【例 2.7】输出真正的 HTML。

```
<div id="app">
    <p>{{website}}</p>
    <p v-html="website"></p>
</div>
<script>
    new Vue({
        el:'#app',
        data:{
             website:'<a class="red" href="https://cn.vuejs.org/">Vue.js官网</a>'
        }
    })
</script>
```

在 IE 11 浏览器中运行，打开控制台，可以发现，使用 v-html 指令的 p 标签输出了真正的 a 标签，当单击“Vue.js 官网”链接后，将跳转到对应的页面，效果如图 2-10 所示。

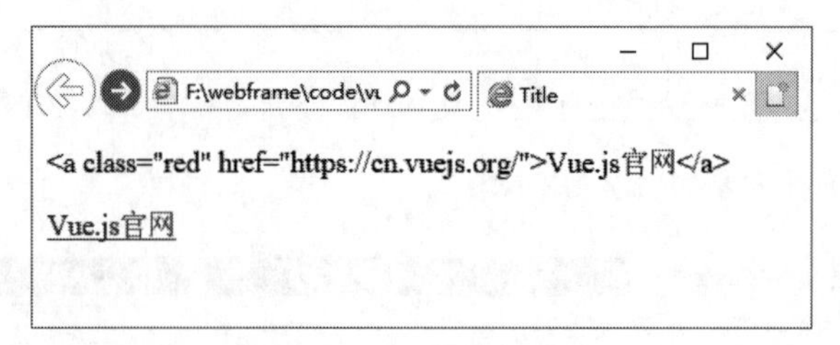

图 2-10　输出真正的 HTML

注意

站点上动态渲染的任意 HTML 可能会非常危险，因为它很容易导致 XSS 攻击。请只对可信内容使用 HTML 插值，绝不要对用户提供的内容使用插值。

3. 使用 JavaScript 表达式

在模板中，一直都只绑定简单的属性键值。但实际上，对于所有的数据绑定，Vue 都提供了完全的 JavaScript 表达式支持。

```
{{ number + 1 }}
{{ ok ? 'YES' : 'NO' }}
{{ message.split('').reverse().join('') }}
<div v-bind:id="'list-' + id"></div>
```

上面这些表达式会在所属 Vue 实例的数据作用域下作为 JavaScript 被解析。限制就是，每个绑定都只能包含单个表达式，所以下面的例子都不会生效。

```
<!-- 这是语句，不是表达式 -->
{{ var a = 1}}
<!-- 流控制也不会生效，请使用三元表达式 -->
{{ if (ok) { return message } }}
```

【例 2.8】使用 JavaScript 表达式。

```
<div id="app">
<p>3条鱼总共{{fish*number+data}}元</p>
</div>
<script>
    new Vue({
        el:"#app",
        data:{
            fish:3,
            number:100,
            data:10
        }
    })
</script>
```

在 IE 11 浏览器里面运行的结果如图 2-11 所示。

图 2-11 使用 JavaScript 表达式计算结果

2.2.2 指令

指令（Directives）是带有“v-”前缀的特殊特性。指令特性的值预期是单个JavaScript表达式（v-for是例外情况）。指令的职责是，当表达式的值改变时，将其产生的连带影响响应式地作用于DOM。

```
<p v-if="boole">现在你看到我了</p>
```

上面代码中，v-if指令将根据表达式布尔值（boole）的真假来插入或移除<p>元素。

1. 参数

一些指令能够接收一个“参数”，在指令名称之后以冒号表示。例如，v-bind指令可以用于响应式地更新HTML特性：

```
<a v-bind:href="url">...</a>
```

在这里href是参数，告知v-bind指令将该元素的href特性与表达式url的值绑定。

v-on指令用于监听DOM事件，例如下面代码：

```
<a v-on:click="doSomething">...</a>
```

其中，参数click是监听的事件名，在后面章节中将会详细介绍v-on指令的具体用法。

2. 修饰符

修饰符（modifier）是以半角句点“.”指明的特殊后缀，用于指出一个指令应该以特殊方式绑定。例如，.prevent修饰符告诉v-on指令对于触发的事件调用event.preventDefault()：

```
<form v-on:submit.prevent="onSubmit">...</form>
```

2.2.3 缩写

“v-”前缀作为一种视觉提示，用来识别模板中Vue特定的特性。在使用Vue为现有标签添加动态行时，“v-”前缀很有帮助。然而，对于一些频繁用到的指令来说，就会感到使用起来很烦琐。同时，在构建由Vue管理所有模板的单页面应用程序(SPA-single page application)时，“v-”前缀也变得没那么重要了。因此，Vue为v-bind和v-on这两个最常用的指令提供了特定简写。

1. v-bind 缩写

```
<!-- 完整语法 -->
<a v-bind:href="url">...</a>
<!-- 缩写 -->
<a :href="url">...</a>
```

2. v-on 缩写

```
<!-- 完整语法 -->
<a v-on:click="doSomething">...</a>
```

```
<!-- 缩写 -->
<a @click="doSomething">...</a>
```

它们看起来可能与普通的HTML略有不同，但“:”与“@”对于特性名来说都是合法字符，在所有支持Vue的浏览器中都能被正确地解析。而且，它们不会出现在最终渲染的标记中。

2.3 疑难解惑

疑问1：对Vue指令中的动态参数的理解。

从Vue 2.6.0版本开始，可以用方括号括起来的JavaScript表达式作为一个指令的参数：

```
<a v-bind:[attributeName]="url"> ... </a>
```

这里的attributeName会被作为一个JavaScript表达式进行动态求值，求得的值将会作为最终的参数来使用。例如，如果Vue实例有一个data属性attributeName，其值为href，那么这个绑定将等价于v-bind:href。

还可以使用动态参数为一个动态的事件名绑定处理函数：

```
<a v-on:[eventName]="doSomething"> ... </a>
```

当eventName的值为focus时，v-on:[eventName]将等价于v-on:focus。

疑问2：Vue中virtual DOM（虚拟DOM）和真实DOM的区别。

虚拟DOM其实就是JavaScript对象，以JavaScript对象的形式去添加DOM元素。

例如，DOM节点在HTML文档中的表现如下：

```
    <ul id="list">
        <li>item 1</li>
        <li>item 2</li>
    </ul>
```

利用JavaScript来编写一个虚拟DOM，在没有渲染的情况下，它就是一个字符串、一个对象。例如下面代码：

```
Let domNode={
        tag:'ul'
        attributes:{id:'list'}
        children:[
        //这里是li
        ]
    }
```

当把信息push到节点中时，才会变成真实DOM。

```
//更新虚拟DOM的代码
    domNode.children.push('<ul>item 3</ul>')
```

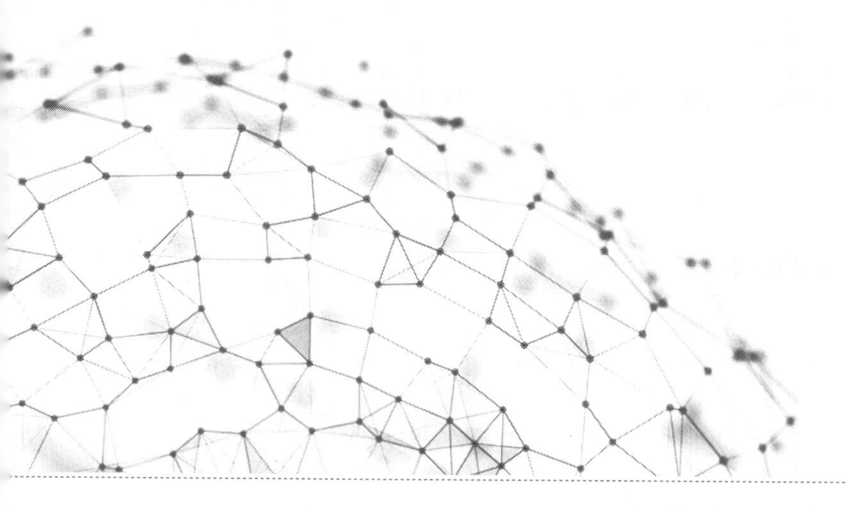

第3章 计算属性、侦听器和过滤器

在 Vue 中，可以很方便地将数据使用插值表达式的方式渲染到页面元素中，但是插值表达式的设计初衷是用于简化运算，不应该对差值做过多的操作。当需要对差值做进一步的处理时，就应该使用 Vue 中的计算属性来完成这一操作。同时，当差值数据变化时，执行异步或开销较大的操作时，可以通过采用监听器的方式来达到目的。

3.1 计算属性

计算属性在 computed 选项中定义。计算属性就是当其依赖属性的值发生变化时，这个属性的值会自动更新，与之相关的 DOM 也会同步更新。这里的依赖属性值是 data 中定义的属性。

下面是一个反转字符串的示例，定义了一个 reversedMessage 计算属性，在 input 输入框中输入字符串时，绑定的 message 属性值发生变化，触发 reversedMessage 计算属性，执行对应的函数，使字符串反转。

【例 3.1】计算属性示例。

```
<div id="app">
    输入内容：<input type="text" v-model="message"><br/>
    反转内容：{{reversedMessage}}
</div>
<script>
    new Vue({
        el: '#app',
        data: {
            message: ''
        },
```

```
        computed: {
            reversedMessage: function () {
                return this.message.split('').reverse().join('')
            }
        }
    })
</script>
```

在 IE 11 浏览器里面运行的结果如图 3-1 所示。

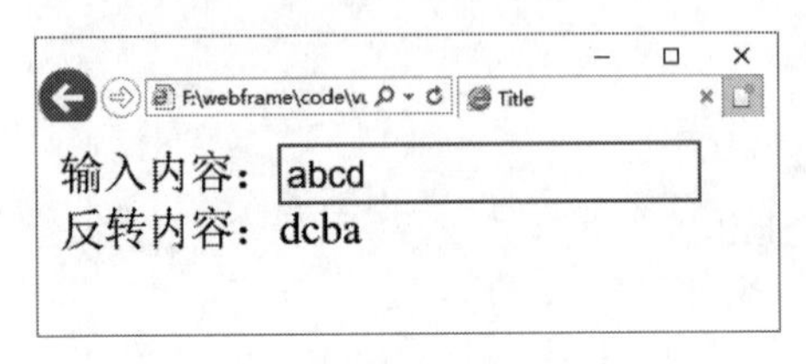

图 3-1　字符串反转效果

3.2 计算属性与方法

计算属性的写法和方法相似，完全可以在 methods 中定义一个方法来实现相同的功能。例如，把上节示例改成用方法实现，代码如下。

【例 3.2】 使用方法实现上节示例。

```
<div id="app">
    输入内容：<input type="text" v-model="message"><br/>
    反转内容：{{reversedMessage()}}
</div>
<script>
    new Vue({
        el: '#app',
        data: {
            message: ''
        },
        //方法
        methods: {
            reversedMessage: function () {
                return this.message.split('').reverse().join('')
            }
        }
    })
</script>
```

在 IE 11 浏览器里面运行效果与前面示例效果完全一样。

其实，计算属性的本质就是一个方法，只不过在使用计算属性的时候，把计算属性的名称直接作为属性来使用，并不会把计算属性作为一个方法去调用。

为什么还要去使用计算属性而不是去定义一个方法呢？计算属性是基于它们的依赖进行缓存的，即只有在相关依赖发生改变时它们才会重新求值。例如，在上面的例子中，只要 message 没有发生改变，多次访问 reversedMessage 计算属性会立即返回之前的计算结果，而不必再次执行函数。

反之，如果使用方法的形式实现，当使用到 reversedMessage 方法时，无论 message 属性是否发生改变，方法都会重新执行一次，这无形中增加了系统的开销。

在某些情况下，计算属性和方法可以实现相同的功能，但有一个重要的不同点：在调用 methods 中的一个方法时，所有方法都会被调用。

例如下面示例，定义了两个方法 add1 和 add2，分别打印“number+a”“number+b”，当调用其中的 add1 时，add2 也将被调用。

【例 3.3】方法的调用。

```
<div id="app">
    <button v-on:click="a++">a+1</button>
    <button v-on:click="b++">b+1</button>
    <p>number+a={{add1()}}</p>
    <p>number+b={{add2()}}</p>
</div>
<script>
     new Vue({
        el: '#app',
        data: {
            a:0,
            b:0,
            number:30
        },
        methods: {
            add1:function(){
                console.log("number+a");
                return this.a+this.number
            },
            add2:function(){
                console.log("number+b")
                return this.b+this.number
            }
        }
    })
</script>
```

在 IE 11 浏览器里面运行，打开控制台，单击“a+1”按钮，可以发现控制台打印了“number+a”和“number+b”，如图 3-2 所示。

图 3-2 方法的调用效果

使用计算属性则不同，计算属性相当于优化了的方法，使用时只会使用对应的计算属性。例如更改上面示例，把 methods 换成 computed，并把 HTML 中的调用 add1 和 add2 方法的括号去掉。

注意

计算属性的调用不能使用括号，例如 add1、add2；调用方法需要加上括号，例如 add1()、add2()。

【例 3.4】计算属性的调用。

```
<div id="app">
    <button v-on:click="a++">a+1</button>
    <button v-on:click="b++">b+1</button>
    <p>number+a={{add1}}</p>
    <p>number+b={{add2}}</p>
</div>
<script>
     new Vue({
        el: '#app',
        data: {
            a:0,
            b:0,
            number:30
        },
        computed: {
            add1:function(){
                console.log("number+a");
                return this.a+this.number
            },
            add2:function(){
                console.log("number+b")
                return this.b+this.number
            }
        }
    })
</script>
```

在 IE 11 浏览器里面运行，打开控制台，在页面中单击“a+1”按钮，可以发现控制台只打印了“number+a”，如图 3-3 所示。

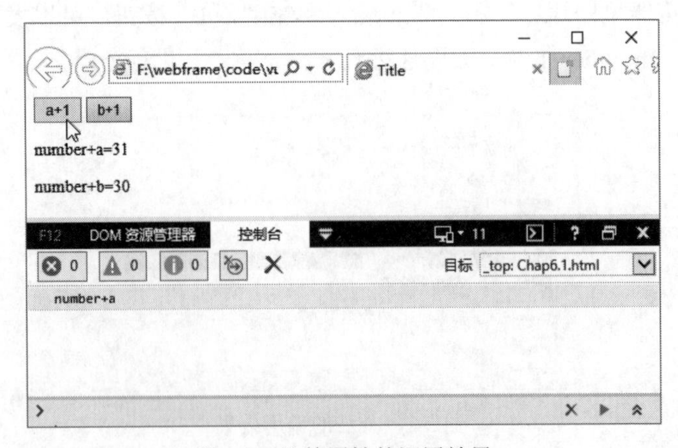

图 3-3　计算属性的调用效果

计算属性相比较于方法更加优化，但并不是什么情况下都使用计算属性，在触发事件时还是使用对应的方法。计算属性一般在数据量比较大、比较耗时的情况下使用（例如搜索），只有虚拟 DOM 与真实 DOM 不同的情况下才会执行 computed。

3.3 侦听属性

在 Vue 中，不光可以使用计算属性的方式来监听数据的变化，还可以使用 watch 监听器的方法来监测某个数据发生的变化。不同的是，计算属性仅仅是对于依赖数据的变化后进行的数据操作，而 watch 更加侧重于对于监测中的某个数据发生变化后所执行的一系列的功能逻辑操作。

监听器以 key-value 的形式定义，key 是一个字符串，它是需要被监测的对象，而 value 则可以是字符串（方法的名称）、函数（可以获取到监听对象改变前的值以及更新后的值）或是一个对象（对象内可以包含回调函数的其他选项，例如是否初始化时执行监听，或是否执行深度遍历，即是否对对象内部的属性进行监听）。

3.3.1 回调值为函数方法

在下面的例子中，我们监听了 message 属性的变化，根据属性的变化执行了回调方法，打印出了属性变化前后的值。

【例 3.5】回调方法。

```
<div id="app">
    输入的值：<input type="text" v-model="message">
</div>
<script>
     new Vue({
        el: '#app',
        data: {
            message: ''
        },
        computed: {},
        watch: {
            message: function (newValue, oldValue) {
                console.log("新值："+newValue+"---------旧值"+oldValue)
            }
        }
    })
</script>
```

在 IE 11 浏览器里面运行，在输入框中输入 1234，控制台的打印结果如图 3-4 所示。

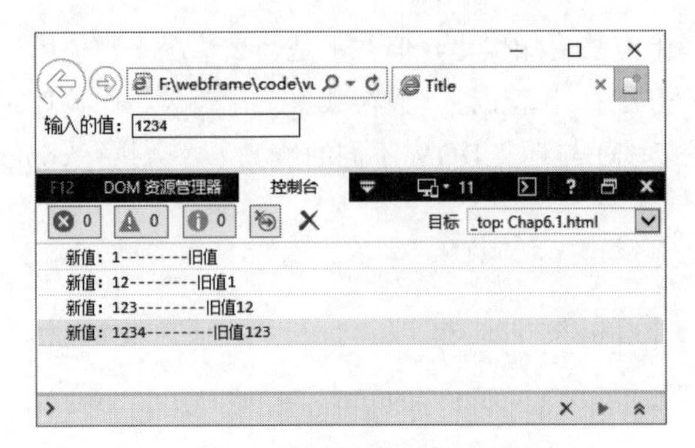

图 3-4　控制台的打印结果

同样，可以通过方法名称指明回调为已经定义好的方法。

【例 3.6】指明回调为已经定义好的方法。

```
<div id="app">
    输入的值：<input type="text" v-model="message">
</div>
<script>
     new Vue({
        el: '#app',
        data: {
            message: ''
        },
        computed: {},
        watch: {
            //调用方法
            message:'way'
        },
            //定义好的方法
         methods:{
            way:function(newValue, oldValue){
                console.log("新值: "+newValue+"--------旧值"+oldValue)
            }
         }
    })
</script>
```

在 IE 11 浏览器里面运行，同样在输入框中输入 1234，效果与上面示例相同。

3.3.2　回调值为对象

当我们监听的回调值为一个对象时，不仅可以设置回调函数，还可以设置一些回调的属性。例如，在下面的例子中，监听了 User 这个对象，同时执行深度遍历，这时监听到 User.name 属性发生改变的时候，就可以执行回调函数。注意，深度遍历默认为 false，当不启用深度遍历时，是无法监听到对象的内部属性的变化的。

【例 3.7】回调值为对象。

```
<div id="app">
    用户姓名：<input type="text" v-model="User.name">
```

```
</div>
<script>
    new Vue({
        el: '#app',
        data: {
            message: '',
            User: {
                name: 'mayun',
            }
        },
        computed: {},
        watch: {
            'User': {
                handler: function (newValue, oldValue) {
                        console.log("对象记录：新值："+newValue.name+"--------- 旧
                          值："+oldValue.name)
                },
                deep: true
            }
        }
    })
</script>
```

在IE 11浏览器里面运行，在“mayun”后面输入1234，控制台的打印效果如图3-5所示。

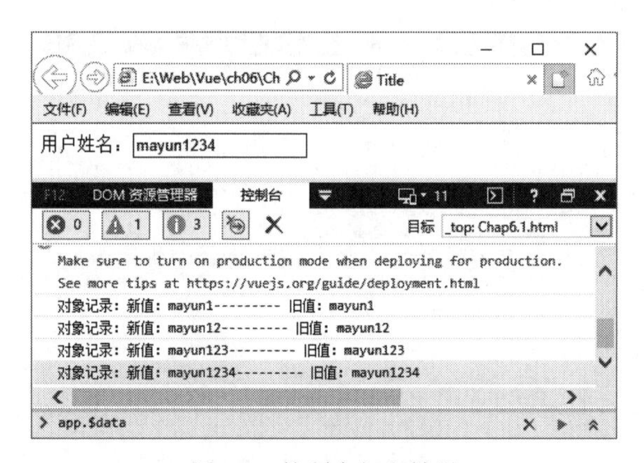

图3-5 控制台打印效果

从上面示例可以发现，newValue与oldValue一样。当监听的数据为对象或数组时，newValue和oldValue是相等的，因为对象和数组都为引用类型，这两个的形参指向的也是同一个数据对象。同时，如果不启用深度遍历，将无法监听到对于User对象中name属性的变化。

【例3.8】不启用深度遍历。

```
<div id="app">
    用户姓名：<input type="text" v-model="User.name">
</div>
<script>
    new Vue({
        el: '#app',
        data: {
```

```
            message: '',
            User: {
                name: 'mayun',
            }
        },
        computed: {},
        watch: {
            //回调为对象
            'User': {
                handler: function (newValue, oldValue) {
                        console.log("对象记录：新值："+newValue.name + "--------- 旧
                          值："+oldValue.name)
                },
                deep: false
            }
        },
        methods: {}
    })
</script>
```

在 IE 11 浏览器里面运行的结果如图 3-6 所示。

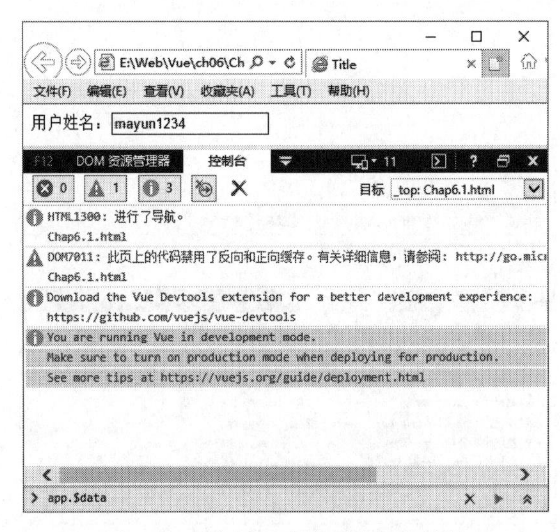

图 3-6　不启用深度遍历的打印结果

从前面内容可以总结：计算属性的结果会被缓存起来，只有依赖的属性发生变化时才会重新计算，必须返回一个数据，主要用来进行纯数据的操作。监听器主要用来监听某个数据的变化，从而去执行某些具体的回调业务逻辑，不仅仅局限于返回数据。

3.4 过滤器

过滤器可对数据进行筛选、过滤、格式化，例如时间格式化、英文大小写转换、状态转换，等等。它与 methods、computed 和 watch 不同的是，其不能改变原始值。

Vue允许自定义过滤器，可被用于一些常见的文本格式化。过滤器可以用在两个地方：双花括号插值和 v-bind 表达式 (从 2.1.0 版开始支持)。过滤器应该被添加在 JavaScript 表达式的尾部，由“管道”符号指示：

```
<!-- 在双花括号中 -->

{{ message | capitalize }}

<!--在v-bind中-->

<div v-bind:id="rawId | formatId"></div>
```

可以在一个组件的选项中定义本地的过滤器：

```
filters: {
  capitalize: function (value) {
    if (!value) return ''
    value = value.toString()
    return value.charAt(0).toUpperCase() + value.slice(1)
  }
}
```

或者在创建 Vue 实例之前全局定义过滤器：

```
Vue.filter('capitalize', function (value) {
  if (!value) return ''
  value = value.toString()
  return value.charAt(0).toUpperCase() + value.slice(1)
})
new Vue({
  // ...
})
```

当全局过滤器和局部过滤器重名时，会采用局部过滤器。

过滤器函数总接收表达式的值 (之前的操作链的结果) 作为第一个参数。在上述例子中，capitalize 过滤器函数将会收到 message 的值作为第一个参数。

过滤器可以串联：

```
{{ message | filterA | filterB }}
```

在这个例子中，filterA 被定义为接收单个参数的过滤器函数，表达式 message 的值将作为参数传入到函数中。然后继续调用同样被定义为接收单个参数的过滤器函数 filterB，将 filterA 的结果传递到 filterB 中。

过滤器是 JavaScript 函数，因此可以接收参数：

```
{{ message | filterA('arg1', arg2) }}
```

这里，filterA 被定义为接收三个参数的过滤器函数。其中 message 的值作为第一个参数，普通字符串 arg1 作为第二个参数，表达式 arg2 的值作为第三个参数。

在下面的示例中，使用过滤器格式化时间。

【例 3.9】格式化时间。

```
<body>
<div id="app">
    <h1>当前时间：{{date|formateTime}} </h1>
</div>
<script>
    //格式化后的时间格式:年-月-日 时：分：秒
    var parseDate = function(datetime){
        return datetime<10?'0'+datetime:datetime;
    }
    var app = new Vue({
        el:'#app',
        data:{
            date:new Date()
        },
        filters:{
            formateTime:function (val) {
                var date = new Date(val);
                var year = date.getFullYear();
                var month = parseDate(date.getMonth()+1);
                var day = parseDate(date.getDate());
                var hours = parseDate(date.getHours());
                var min = parseDate(date.getMinutes());
                var sec = parseDate(date.getSeconds());
                return year +'-'+month +'-'+day+' '+hours +":"+min+":"+sec;
            }
        },
        created:function(){
            var that = this; //作用域一致
            this.timer = setInterval(function () {
                that.date = new Date();
            },1000);
        },
        beforeDestroy:function(){
            if (this.timer) {
                clearInterval(this.timer);
            }
        }
    })
</script>
</body>
```

在 IE 11 浏览器里面运行的结果如图 3-7 所示。

图 3-7 格式化时间

3.5 疑难解惑

疑问 1：在 Vue 中，可以通过一个计算属性和侦听属性来监听某个数据属性的变化。当一个数据属性在它所依赖的属性发生变化时也要发生变化，这种情况，选择计算属性还是侦听属性呢？

建议选择计算属性。

例如，在下面这个例子中，如果使用监听函数，代码就会变得有点冗余。

```
<div id="app">
    {{ fullName }}
</div>
<script>
    var vm = new Vue({
        el: '#app',
        data: {
            firstName: '杜',
            lastName: '鹃',
            fullName: '杜鹃'
        },
        watch: {
            firstName: function (val) {
                this.fullName = val + ' ' + this.lastName
            },
            lastName: function (val) {
                this.fullName = this.firstName + ' ' + val
            }
        }
    })
</script>
```

fullName 属性依赖于 firstName 和 lastName，这里有个缺点就是，无论 firstName 或 lastName 其中的任何一个发生改变时，都要调用不同的监听函数来更新 fullName 属性。

但是当使用计算属性时，代码就会变得更加简洁，代码如下：

```
<div id="app">
    {{ fullName }}
</div>
<script>
    var vm = new Vue({
        el: '#app',
        data: {
            firstName: '杜',
            lastName: '鹃'
        },
        computed: {
            fullName: function () {
                return this.firstName + ' ' + this.lastName
            }
        }
    })
</script>
```

疑问 2：计算属性中 setter 属性的用法？

计算属性默认只有 getter，不过在需要时也可以使用 setter。默认情况下是使用 getter 读取：

```
<div id="app">
    {{ fullName }}
</div>
<script>
    var vm = new Vue({
        el: '#app',
        data: {
            firstName: '杜',
            lastName: '鹃'
        },
        computed: {
            fullName: {
                //getter
                get: function () {
                    return this.firstName + ' ' + this.lastName
                }
            }
        }
    });
</script>
```

当手动修改计算属性的值，像修改一个普通数据那样时，就会触发 setter，执行一些自定义的操作，例如下面的代码：

```
…
computed: {
    fullName: {
        //getter
        get: function () {
            return this.firstName + ' ' + this.lastName
        },
        //setter
        set: function (newValue) {
            var names = newValue.split(' ');
            this.firstName = names[0];
            this.lastName = names[names.length - 1];
        }
    }
}
…
```

当运行 vm.fullName='杜甫'时，setter 会被调用，vm.firstName 和 vm.lastName 也会相应地被更新。

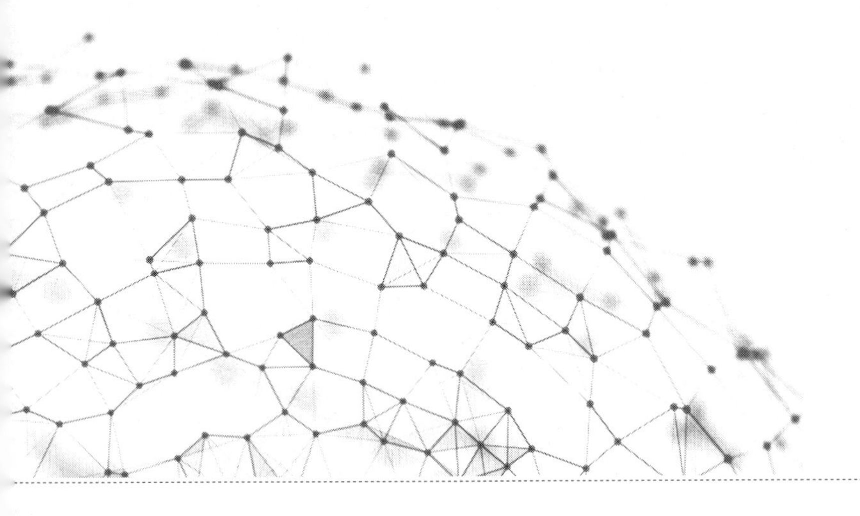

第4章 内置指令

指令是 Vue 模板中最常用的一项功能，它带有前缀 v-，主要职责是当其表达式的值改变时，相应地将某些行为应用在 DOM 上。本章开始学习基本指令、条件渲染、列表渲染和自定义指令。

4.1 基本指令

Vue 的各种指令（Directives）更加方便去实现数据驱动 DOM，其中一些基本指令包括 v-cloak、v-bind、v-on、v-model、v-once、v-text 和 v-html 等。

4.1.1 v-cloak

在使用 Vue 的过程中，当引入了 vue.js 这个文件之后，浏览器的内存中就存在了一个 Vue 对象，我们可以通过构造函数的方式创建出一个 Vue 的对象实例，后面就可以对该实例进行操作。

如果在这个过程中，对于 vue.js 的引入因为某些原因而没有加载完成，此时，未编译的 Mustache 标签就无法正常显示。例如，在下面的例子中，我们模拟将网页加载速度变慢，此时就可以看见，页面最先开始会显示出插值表达式，只有 vue.js 加载完成后，才会渲染成正确的数据。

【例 4.1】 v-cloak 示例。

```
<div id="app">
    <p>{{message}}</p>
</div>
```

```
<script>
    new Vue({
        el: '#app',
        data: {
            message: 'hello world!'
        }
    });
</script>
```

在 IE 11 浏览器里面运行，可以发现，vue.js 加载完成后，才会渲染成正确的数据。

这时虽然已经加了指令 v-cloak，但其实并没有起到任何作用，当网速较慢、vue.js 文件还没加载完时，在页面上会显示 {{message}} 的字样，直到 Vue 创建实例、编译模板时，DOM 才会被替换，所以这个过程屏幕是有闪动的。只要加上下面一段 CSS 就可以解决这个问题：

```
<!-- 添加 v-cloak 样式 -->
<style>
    [v-cloak] {
        display: none;
    }
</style>
```

4.1.2 v-once

v-once 指令只渲染元素和组件一次，随后的渲染，使用了此指令的元素、组件及其所有的子节点，都会当作静态内容并跳过，这可以用于优化更新性能。

例如，在下面的示例中，当修改 input 输入框的值时，使用了 v-once 指令的 p 元素，不会随着输入框的内容改变，而第二个 p 元素跟随着输入框的内容改变。

【例 4.2】v-once 示例。

```
<div id="app">
    <p v-once>不可改变：{{msg}}</p>
    <p>可以改变：{{msg}}</p>
    <p><input type="text" v-model = "msg" name=""></p>
</div>
<script>
    new Vue({
        el: '#app',
        data: {
            msg : "hello"
        }
    })
</script>
```

在 IE 11 浏览器里面运行，然后在输入框中输入 123，可以看到，添加 v-once 指令的 p 标签，并没有任何的变化，效果如图 4-1 所示。

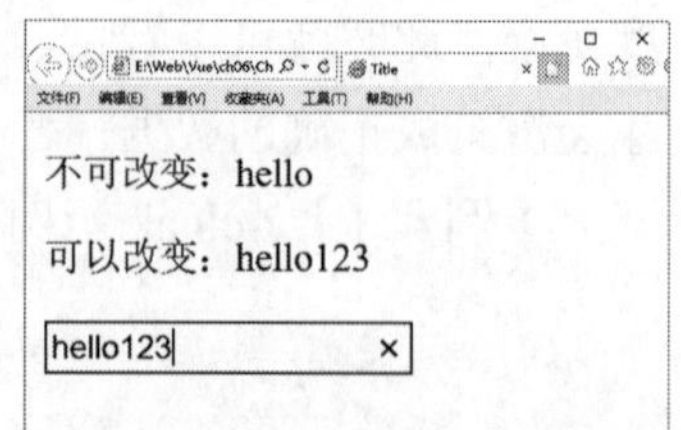

图 4-1　v-once 指令作用效果

4.1.3 v-text 与 v-html

v-text 与 v-html 指令都可以更新页面元素的内容，不同的是，v-text 会将数据以字符串文本的形式更新，而 v-html 则是将数据以 HTML 标签的形式更新。

在更新数据上，我们也可以使用差值表达式进行数据更新，不同于 v-text、v-html 指令，差值表达式只会更新原本占位插值所在的数据内容，而 v-text、v-html 指令则会替换掉整个内容。

【例 4.3】v-text 与 v-html 指令示例。

```
<div id="app">
    <p>******{{message}}******</p>
    <p v-text="message">************</p>
    <p v-text="Html">************</p>
    <p v-html="Html">************</p>
</div>
<script>
    new Vue({
        el: '#app',
        data: {
            message: 'hello world!',
            Html: '<h3 style="color:blue">Vue是现在最流行的框架之一</h3>'
        }
    });
</script>
```

在 IE 11 浏览器里面运行的结果如图 4-2 所示。

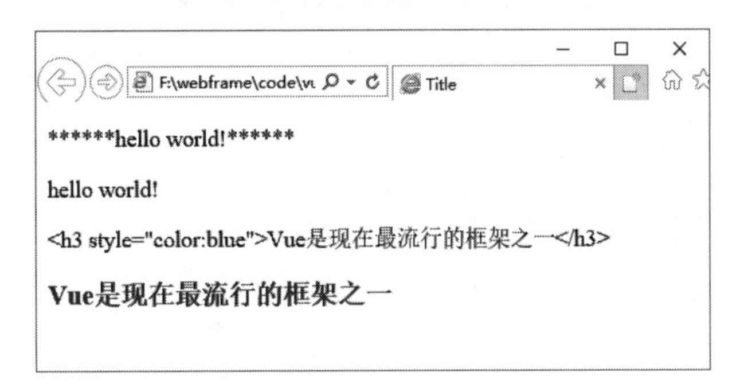

图 4-2　v-text 与 v-html 指令作用效果

4.1.4 v-bind

v-bind 可以用来在标签上绑定标签的属性（例如：img 的 src、title 属性等）和样式（可以用 style 的形式进行内联样式的绑定，也可以通过指定 class 的形式指定样式）。同时，对于绑定的内容，是作为一个 JavaScript 变量，因此，可以对该内容进行编写合法的 JavaScript 表达式。

在下面的示例中，按钮的 title 和 style 都是通过 v-bind 指令进行绑定的，这里对于样式的绑定，需要构建一个对象。其他对于样式的绑定方法，将在后面的学习中提到。

【例 4.4】v-bind 指令示例。

```
<div id="app">
<!--v-bind：可以用来在标签上绑定标签的属性和样式，对于绑定的内容，可以对该内容进行编写合法的 JavaScript表达式-->
<input type="button" value="按钮" v-bind:title="Title" v-bind:style="{color:Color,width:Width+'px'}">
<!-- 简写形式如下：-->
<!--<input type="button" value="按钮" :title="Title" :style="{color:Color,width:Width+'px'}">-->
</div>
<script>
    new Vue({
        el: '#app',
        data: {
            Title: '这是我自定义的title属性',
            Color: 'red',
            Width: '120'
        }
    });
</script>
```

在IE 11浏览器里面运行，打开控制台，可以看到数据已经渲染到了DOM中，如图4-3所示。

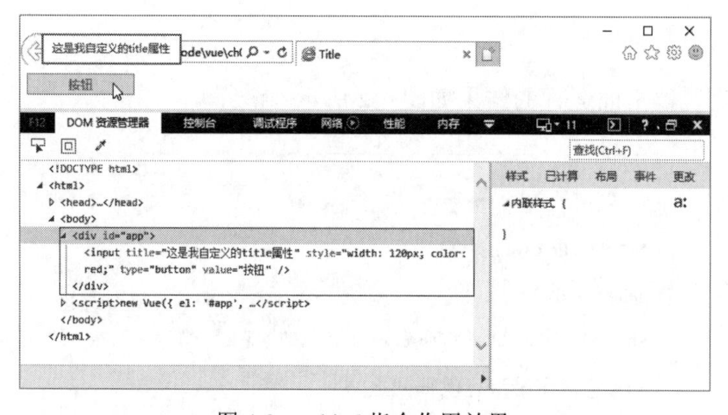

图 4-3　v-bind 指令作用效果

4.1.5　v-on

在传统的前端开发中，想对一个按钮绑定事件时，需要获取到这个按钮的DOM元素，再对这个获取到的元素进行事件的绑定。在 Vue 中，对于 DOM 的操作，全部让 Vue 替我们完成，我们只关注业务代码实现，使用 Vue 内置的 v-on 指令来完成事件的绑定。

【例 4.5】传统的 JavaScript 实现单击事件。

```
<div id="app">
    <input type="button" value="单击事件" id="btn">
</div>
<script>
    //传统的事件绑定方法
```

```
    document.getElementById('btn').onclick = function () {
        alert('传统的事件绑定方法');
    }
</script>
```

在 IE 11 浏览器里面运行，单击“单击事件”按钮，触发事件，效果如图 4-4 所示。

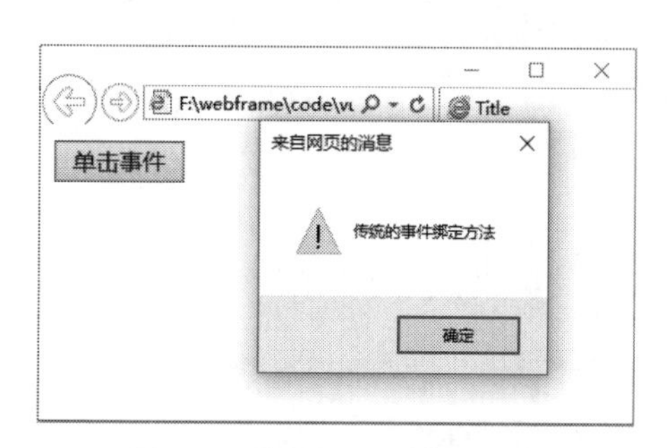

图 4-4　传统的单击事件效果

在使用 v-on 指令对事件进行绑定时，需要在标签上指明 v-on:event(click、mousedown、mouseup 等) 绑定的事件。

【例 4.6】v-on 指令示例。

```
<div id="app">
    <input type="button" value="单击事件" v-on:click="alert('Vue中的事件绑定')">
</div>
<script>
    new Vue({
        el: '#app'
    });
</script>
```

在 IE 11 浏览器里面运行，单击“单击事件”按钮，触发事件，效果如图 4-5 所示。

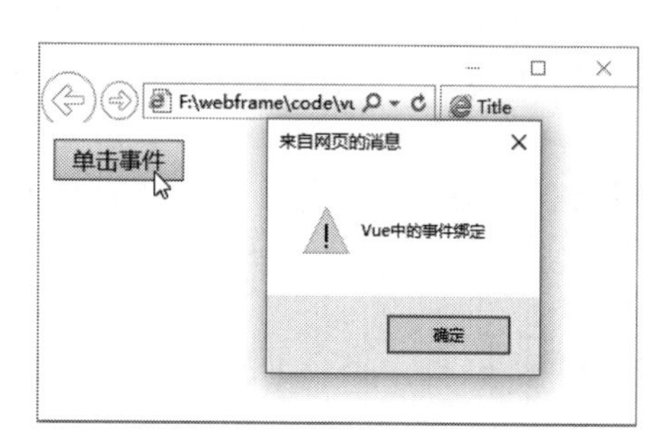

图 4-5　v-on 指令作用效果

在 Vue 的设计中许多事件处理逻辑会更为复杂，所以直接把 JavaScript 代码写在 v-on 指令中是不可行的。因此 v-on 还可以接收一个需要调用的方法名称，可以在 Vue 实例的 methods 属性下写出方法，在方法中编写逻辑代码。

【例 4.7】methods 属性示例。

```
<div id="app">
    <input type="button" value="单击事件" v-on:click="Alert()">
```

```
</div>
<script>
    new Vue({
        el: '#app',
        //在methods属性中定义Alert方法
        methods:{
            Alert:function(){
                alert("Vue中的事件绑定")
            }
        }
    });
</script>
```

在 IE 11 浏览器里面运行的结果如图 4-6 所示。

图 4-6　methods 属性实现效果

使用 v-on 指令接收的方法名称也可以传递参数，只需要在 methods 中定义方法时说明这个形参即可。

4.2 条件渲染

v-if、v-show 可以实现条件渲染，Vue 会根据表达式值的真假条件来渲染元素。还有可以与 v-if 搭配的 v-else、v-else-if 指令，类似于 JavaScript 中的 if-else、if-elseif-elseif。

4.2.1　v-if

v-if 指令用于条件性地渲染一块内容。这块内容只会在指令的表达式返回 truthy 值的时候被渲染。

提示

在 JavaScript 中，truthy(真值)指的是在布尔值上下文中转换后的值为真的值。所有值都是真值，除非它们被定义为 falsy(即除了 false、0、""、null、undefined 和 NaN 外)。

【例 4.8】v-if 指令示例。

```
<div id="app">
    马云正在学习
    <h3 v-if="value">Vue.js</h3>
    <h3 v-if="!value">Angular.js</h3>
</div>
<script>
    new Vue({
        el: '#app',
        data: {
            value:true,
        }
    })
</script>
```

在 IE 11 浏览器里面运行的结果如图 4-7 所示。

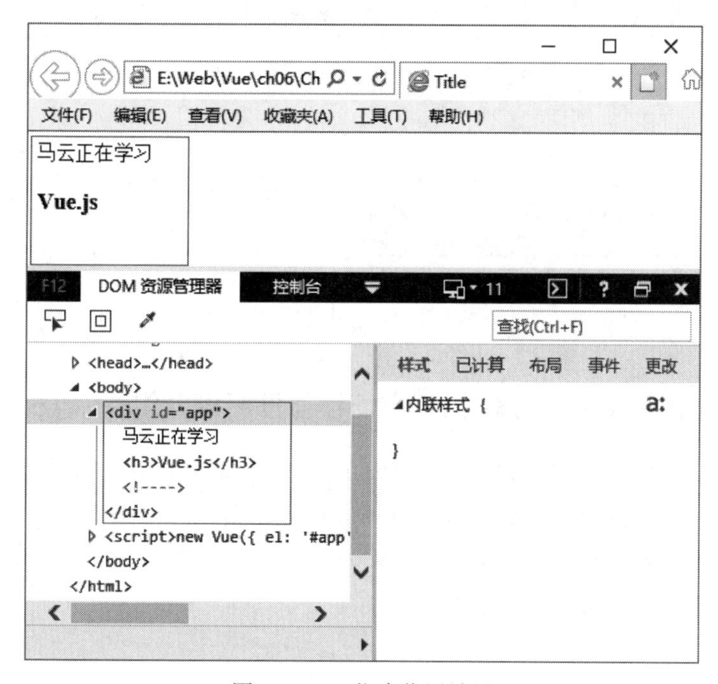

图 4-7　v-if 指令作用效果

在上面示例中，value 值为 true，!true 的值则为 false。也可以用 v-else 添加一个“else 块”。

【例 4.9】添加一个“else 块”。

```
<div id="app">
    马云正在学习
    <h3 v-if="!value">Vue.js</h3>
    <h3 v-else>Angular.js</h3>
</div>
<script>
    new Vue({
        el: '#app',
        data: {
```

```
            value:true,
        }
    })
</script>
```

在 IE 11 浏览器里面运行的结果如图 4-8 所示。

图 4-8 “else 块”作用效果

4.2.2 在 <template> 元素上使用 v-if 条件渲染分组

因为 v-if 是一个指令，所以必须将它添加到一个元素上。但是如果想切换多个元素，此时可以把一个 <template> 元素当作不可见的包裹元素，并在上面使用 v-if，最终的渲染结果将不包含 <template> 元素。

【例 4.10】使用 <template> 包裹元素。

```
<div id="app">
    <template v-if="value">
        <h3>马云想要学习</h3>
        <p>Vue.js</p>
        <p>Angular.js</p>
    </template>
</div>
<script>
    new Vue({
        el: '#app',
        data: {
            value:true,
        }
    })
</script>
```

在 IE 11 浏览器里面运行，打开控制台，可以看到最终的渲染结果将不包含<template>元素，如图 4-9 所示。

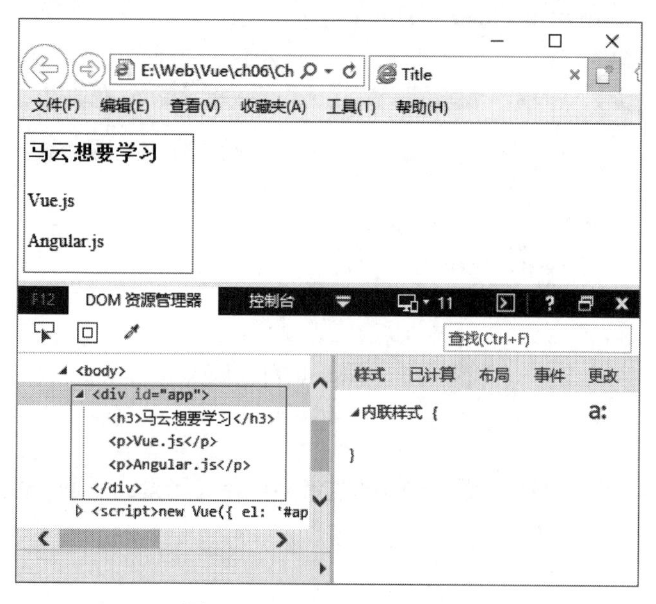

图 4-9　<template>包裹元素

4.2.3　v-else

可以使用 v-else 指令来表示 v-if 的“else 块”，类似于 JavaScript 中的 if-else 逻辑语句。

【例 4.11】v-else 指令示例。

```
<div id="app">
    <h3>value此时的值:{{value}}</h3>
    <!--如果value>0.5-->
    <div v-if="value>0.5">
        你现在可以看到我
    </div>
    <!--否则-->
    <div v-else>
        你现在看不到我
    </div>
</div>
<script>
    new Vue({
        el: '#app',
        data: {
            value:Math.random()   //定义一个随机值
        }
    })
</script>
```

在 IE 11 浏览器里面运行的结果如图 4-10 所示。

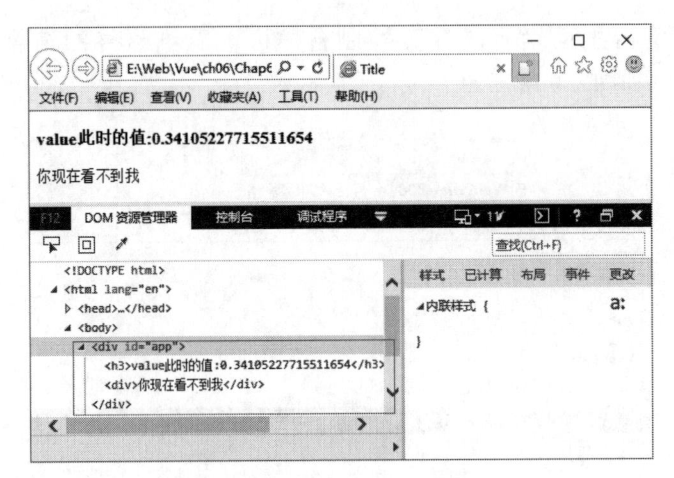

图 4-10　v-else 指令作用效果

注意

v-else 元素必须紧跟在带 v-if 或者 v-else-if 元素的后面，否则将不会被识别。

4.2.4　v-else-if

v-else-if 指令类似于条件语句中的“else-if 块”，可以与 v-if 连续使用。

【例 4.12】 v-else-if 指令示例。

```
<div id="app">
    <div v-if="type === 'A'">
        A
    </div>
    <div v-else-if="type === 'B'">
        B
    </div>
    <div v-else-if="type === 'C'">
        C
    </div>
    <div v-else>
        Not A/B/C
    </div>
</div>
<script>
    new Vue({
        el: '#app',
        data: {
            type:"E"
        }
    })
</script>
```

在 IE 11 浏览器里面运行的结果如图 4-11 所示。

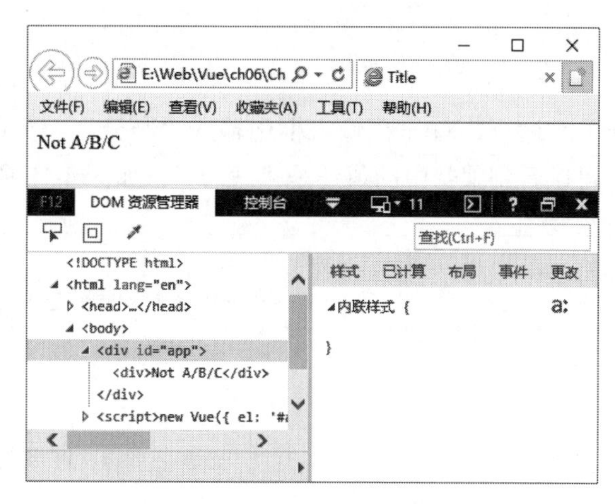

图 4-11 v-else-if 指令作用效果

注意

类似于 v-else，v-else-if 也必须紧跟在带 v-if 或者 v-else-if 的元素之后。

4.2.5 用 key 管理可复用的元素

Vue 会尽可能高效地渲染元素，通常会复用已有元素而不是从头开始渲染。这么做除了使 Vue 变得非常快之外，还有其他一些好处。例如，允许用户在不同的登录方式之间切换。

【例 4.13】 不添加 key 属性示例。

```
<div id="app">
    <template v-if="loginType === 'username'">
        <label>姓名：</label>
        <input placeholder="输入用户名">
    </template>
    <template v-else>
        <label>邮箱：</label>
        <input placeholder="输入邮箱">
    </template>
    <button @click="toggleLoginType">切换</button>
</div>
<script>
    new Vue({
        el: '#app',
        data: {
            loginType: 'username'
        },
        methods: {
            toggleLoginType: function() {
                return this.loginType = this.loginType === 'username' ? 'email'
                 : 'username'
            }
        }
```

```
    });
</script>
```

在 IE 11 浏览器里面运行，首先在输入框中输入“马云”，如图 4-12 所示；然后单击“切换”按钮，可以看到邮箱中的值还是“马云”，如图 4-13 所示。

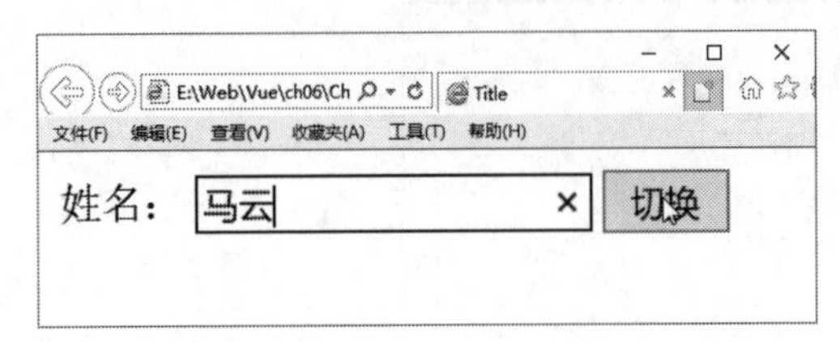

图 4-12　输入“马云”

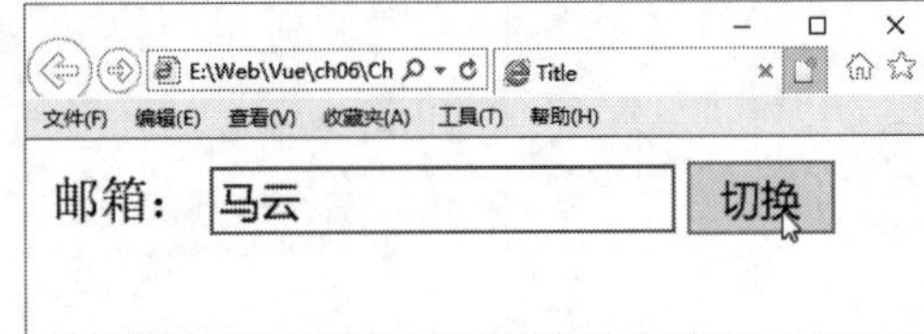

图 4-13　切换效果

在上面的示例中，切换将不会清除用户已经输入的内容。因为两个模板使用了相同的元素，<input> 不会被替换掉，仅仅是替换了它的 placeholder。

这样也不总是符合实际需求，所以 Vue 提供了一种方式来表达“这两个元素是完全独立的，不要复用它们”。只需添加一个具有唯一值的 key 属性即可。

【例 4.14】添加 key 属性示例。

```
<div id="app">
    <template v-if="loginType === 'username'">
        <label>姓名：</label>
        <input placeholder="输入用户名" key="value1">
    </template>
    <template v-else>
        <label>邮箱：</label>
        <input placeholder="输入邮箱" key="value2">
    </template>
    <button @click="toggleLoginType">切换</button>
</div>
<script>
    new Vue({
        el: '#app',
        data: {
            loginType: 'username'
        },
        methods: {
            toggleLoginType: function() {
                return this.loginType = this.loginType === 'username' ? 'email'
                 : 'username'
            }
        }
    });
</script>
```

在 IE 11 浏览器里面运行，首先在输入框中输入“马云”，如图 4-14 所示；然后单击“切换”按钮，可以看到邮箱中的值为默认的 placeholder，如图 4-15 所示。

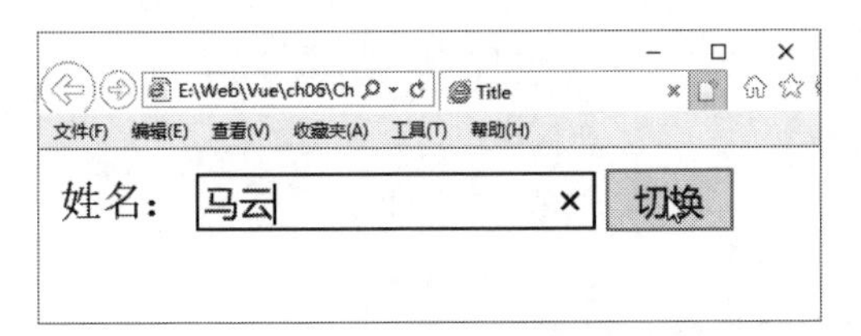

图 4-14 输入“马云”

图 4-15 切换效果

注意

<label> 元素仍然会被高效地复用，因为它们没有添加 key 属性。

4.2.6 v-show

根据条件展示元素的选项还有 v-show 指令，用法与 v-if 大致一样，不同的是带有 v-show 的元素始终会被渲染并保留在 DOM 中。

v-show 只是简单地切换元素的 CSS 属性 display，当模板属性为 true 的时候，控制台显示为 display:block；属性值为 false 的时候，控制台显示 display: none。

注意

v-show 不支持 <template> 语法，也不支持 v-else。

【例 4.15】v-show 指令示例。

```
<div id="app">
    马云正在学习
    <h3 v-show="value">Vue.js</h3>
    <h3 v-show="!value">Angular.js</h3>
</div>
<script>
    new Vue({
        el: '#app',
        data: {
            value:true,
        }
    })
</script>
```

在 IE 11 浏览器里面运行的结果如图 4-16 所示。

图 4-16 v-show 指令作用效果

4.2.7 v-if 与 v-show 的区别

v-if 与 v-show 指令都是根据表达式的真假值判断元素的显示与隐藏。

v-if 是“真正”的条件渲染，因为它会确保在切换过程中，条件块内的事件监听器和子组件适当地被销毁和重建。

v-if 也是惰性的：如果在初始渲染时条件为假，则什么也不做，直到条件第一次变为真时，才会开始渲染条件块。

相比之下，v-show 就简单得多，不管初始条件是什么，元素总是会被渲染，并且只是简单地基于 CSS 进行切换。

一般来说，v-if 有更高的切换开销，而 v-show 有更高的初始渲染开销。因此，如果需要非常频繁地切换，则使用 v-show 较好；如果在运行时条件很少改变，则使用 v-if 较好。

在下面的示例中，我们使用 v-if 与 v-show 指令，通过绑定一个按钮的单击事件，去修改 message 值，从而做到对于两张图片标签的显示与隐藏的控制。

提示

其中，this.message=!this.message; 作用是动态切换 message 的值（true），当单击按钮时，message 的值变为相反的值（false）。

【例 4.16】v-if 与 v-show 比较示例。

```
<div id="app">
    <input type="button" value="切换" @click="Click"><br/>
    <!--v-if 指令控制-->
    <img v-if="message" src="img/001.png" alt="" width="200">
    <!-- v-show 指令控制-->
    <img v-show="message" src="img/002.png" alt="" width="200">
</div>
<script>
     new Vue({
        el: '#app',
        data: {
```

```
                message: true
            },
            methods: {
                Click:function() {
                    this.message=!this.message;
                }
            }
        });
    </script>
```

在 IE 11 浏览器里面运行，由于 message 的值为 true，所以两张图片显示效果如图 4-17 所示。

图 4-17　页面初始化效果

当单击“切换”按钮后，message 的值变成 false，两张图片都被隐藏，效果如图 4-18 所示。

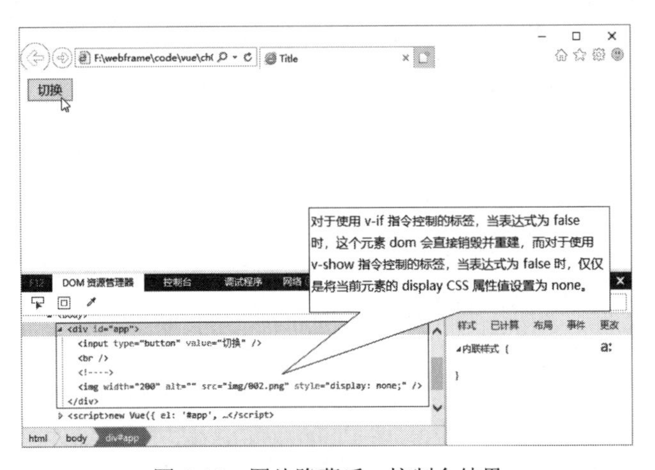

图 4-18　图片隐藏后，控制台结果

从上面的示例可以看到，message 的初始值为 true，此时，两个图片标签都可以显示出来，当单击“切换”按钮后，两张图片都隐藏。不同的是，我们可以看到，使用 v-if

指令控制的 img 标签，当表达式为 false 时，这个元素 DOM 会直接销毁并重建，而使用 v-show 指令控制的 img 标签，当表达式为 false 时，仅仅是将当前元素的 display CSS 属性值设置为 none。所以，当需要频繁控制元素的显示与隐藏时，推荐使用 v-show 指令，避免因为使用 v-if 指令而造成的较高的资源消耗。

4.3 列表渲染

当遍历一个数组或枚举一个对象进行迭代循环展示时，就会用到列表渲染指令 v-for。

4.3.1 使用 v-for 指令遍历元素

不管是写 C#、java 还是前端的 JavaScript 脚本，提到循环数据，首先都会想到使用 for 循环。同样，在 Vue 中，也为我们提供了 v-for 指令用来循环数据。

在使用 v-for 指令时，可以对数组、对象、数字、字符串进行循环，来获取源数据中的每一个值。使用 v-for 指令，必须使用特定语法，其中 items 是源数据数组，而 item 则是被迭代的数组元素的别名，具体格式如下：

```
<div v-for="item in items">
  {{ item.text }}
</div>
```

1. 使用 v-for 遍历数组

【例 4.17】遍历数组示例。

```
<div id="app">
    <h3>评选出今年最火的几个人：</h3>
    <ul>
        <li v-for="item in items">
            {{ item }}
        </li>
    </ul>
</div>
<script>
    new Vue({
        el: '#app',
        data: {
            items: ['马云','马化腾','蔡徐坤','刘强东']
        }
    })
</script>
```

在 IE 11 浏览器里面运行的结果如图 4-19 所示。

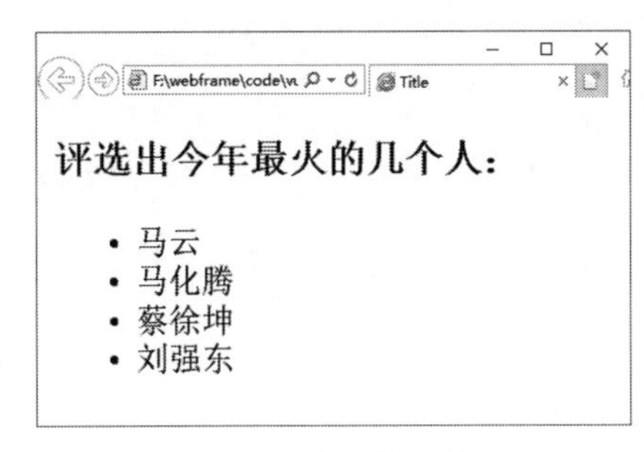

图 4-19　遍历数组效果

在 v-for 指令中，可以访问所有父作用域的属性。v-for 还支持一个可选的第二参数，即当前项的索引，案例如下。

【例 4.18】v-for 指令第二参数。

```
<div id="app">
    <h3>评选出今年最火的几个人：</h3>
    <ul>
        <li v-for="(item,index) in items">
            {{ index }}-{{ item }}
        </li>
    </ul>
</div>
```

在 IE 11 浏览器里面运行的结果如图 4-20 所示。

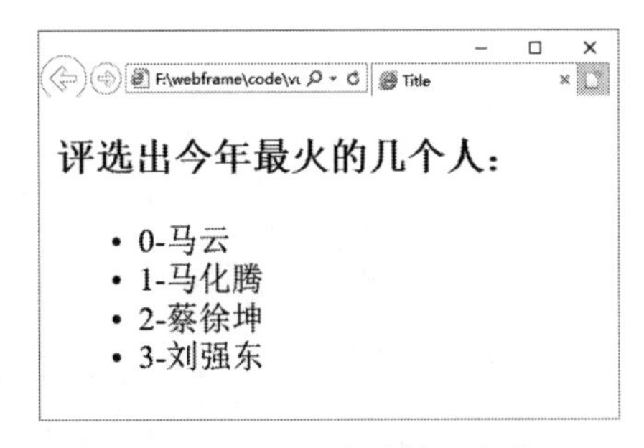

图 4-20　v-for 指令第二参数

提示

可以使用 of 替代 in 作为分隔符，因为它更接近 JavaScript 迭代器的语法：

```
<div v-for="item of items"></div>
```

2. 使用 v-for 遍历对象

【例 4.19】遍历对象。

```
<div id="app">
    <h3>人物介绍</h3>
    <ul>
        <li v-for="value in object">
            {{ value }}
```

```
            </li>
        </ul>
</div>
<script>
    new Vue({
        el: '#app',
        data: {
            //定义对象
            object: {
                姓名: '蔡徐坤',
                性别: '男',
                出生日期: '2019-10-10'
            }
        }
    })
</script>
```

在 IE 11 浏览器里面运行的结果如图 4-21 所示。

也可以提供第二参数为 property 名称 (也就是键名):

```
<div id="app">
    <h3>人物介绍</h3>
    <ul>
        <li v-for="(value,name) in object">
            {{name}}: {{ value }}
        </li>
    </ul>
</div>
```

在 IE 11 浏览器中运行，效果如图 4-22 所示。

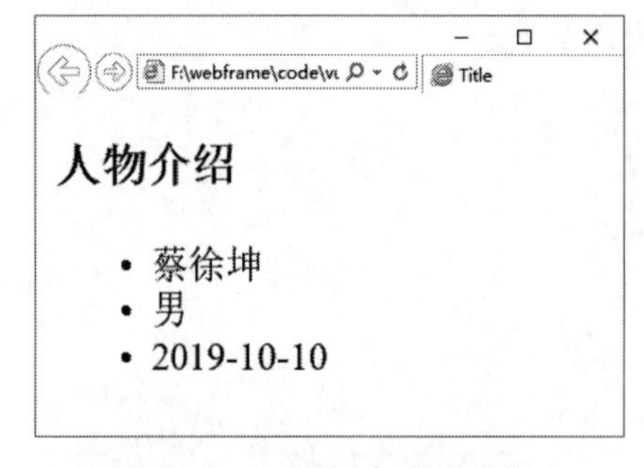

图 4-21　v-for 指令遍历对象

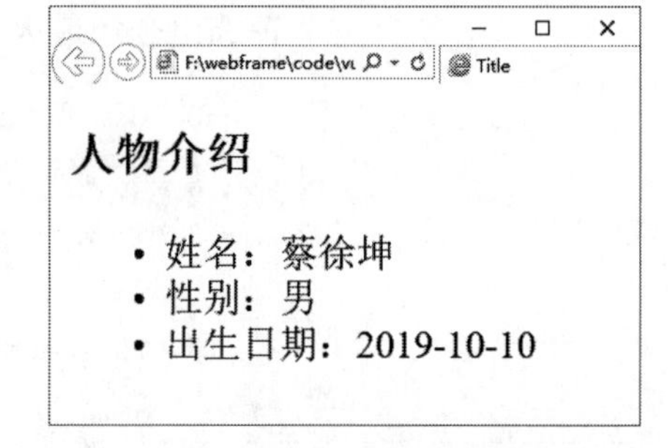

图 4-22　v-for 的第二参数（键名）

还可以用第三参数作为索引：

```
<div id="app">
    <h3>人物介绍</h3>
    <ul>
        <li v-for="(value,name,index) in object">
            {{index}}-{{name}}: {{ value }}
        </li>
    </ul>
</div>
```

在 IE 11 浏览器中运行，效果如图 4-23 所示。

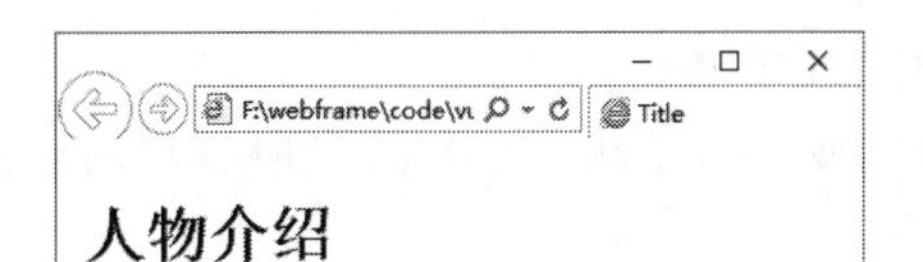

图 4-23　v-for 的第三参数（索引）

注意

在遍历对象时，会按 Object.keys() 的结果遍历，但是不能保证它的结果在不同的 JavaScript 引擎下都一致。Object.keys() 用来获取对象自身可枚举的属性键。

4.3.2　维护状态

当 Vue 正在更新使用 v-for 渲染的元素列表时，它默认使用“就地更新”的策略。如果数据项的顺序被改变，Vue 将不会移动 DOM 元素来匹配数据项的顺序，而是就地更新每个元素，并且确保它们在每个索引位置正确渲染。这个类似 Vue 1.x 的 track-by="$index"。

这个默认的模式是高效的，但是只适用于不依赖子组件状态或临时 DOM 状态 (例如：表单输入值) 的列表渲染输出。

为了给 Vue 一个提示，以便它能跟踪每个节点的身份，从而重用和重新排序现有元素，需要为每项提供一个唯一 key 属性：

```
<div v-for="item in items" v-bind:key="item.id">
    <!--内容-->
</div>
```

建议尽可能在使用 v-for 时提供 key 属性，除非遍历输出的 DOM 内容非常简单或者是刻意依赖默认行为以获取性能上的提升。

注意

不要使用对象或数组之类的非基本类型值作为 v-for 的 key。请用字符串或数值类型的值。

4.3.3　数组更新检测

Vue 为了增强列表渲染的功能，增加了一组观察数组的方法，而且可以显示一个数组的过滤或排序的副本。

1. 变异方法 (mutation method)

Vue 将被侦听的数组的变异方法进行了包裹，所以它们也将会触发视图更新。这些被包裹过的方法如下。

- push()：接收任意数量的参数，把它们逐个添加到数组末尾，并返回修改后数组的长度。
- pop()：从数组末尾移除最后一项，减少数组的 length 值，然后返回移除的项。
- shift()：移除数组中的第一个项并返回该项，同时数组的长度减 1。
- unshift()：在数组前端添加任意个项并返回新数组长度。
- splice()：删除原数组的一部分成员，并可以在被删除的位置加入新的数组成员。
- sort()：调用每个数组项的 toString() 方法，然后比较得到的字符串排序，返回经过排序之后的数组。
- reverse()：用于反转数组的顺序，返回经过排序之后的数组。

这些方法类似于 JavaScript 中操作数组的方法，下面使用 push() 方法进行演示。

【例 4.20】 push() 方法示例。

```
<div id="app">
    <h3>评选出今年最火的几个人：</h3>
    <ul>
    <li v-for="item in items">
        {{ item }}
    </li>
    </ul>
</div>
<script>
    var app=new Vue({
        el: '#app',
        data: {
            items: ['马云','马化腾','蔡徐坤','刘强东']
        }
    })
    //使用push()方法
    app.items.push('小明');
</script>
```

在 IE 11 浏览器里面运行，可以看到在数组的最后渲染了“小明”，结果如图 4-24 所示。

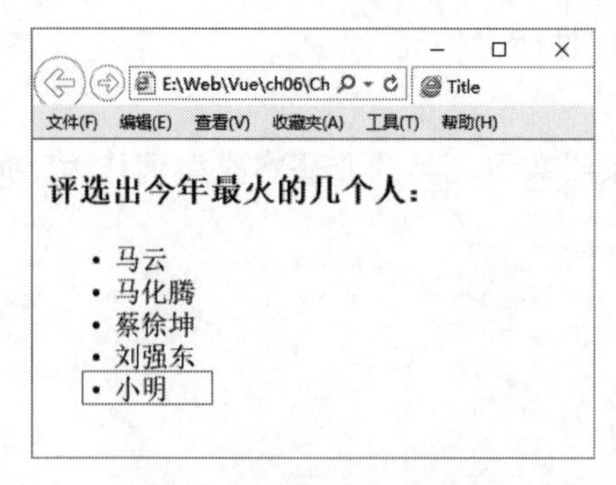

图 4-24　push() 方法示例

2. 替换数组

变异方法会改变调用了这些方法的原始数组。相比之下，也有非变异方法，例如filter()、concat()和slice()。它们不会改变原始数组，而总是返回一个新数组。当使用非变异方法时，可以用新数组替换旧数组。非变异方法如下。

- concat()：先创建当前数组一个副本，然后将接收到的参数添加到这个副本的末尾，最后返回新构建的数组。
- slice()：基于当前数组中一个或多个项创建一个新数组，接受一个或两个参数，即要返回项的起始和结束位置，最后返回新数组。
- map()：对数组的每一项运行给定函数，返回每次函数调用的结果组成的数组。
- filter()：对数组中的每一项运行给定函数，该函数会返回true的项组成的数组。

例如，要显示一个数组的过滤或排序副本，而不实际改变或重置原始数据（非变异方法），可以使用filter()方法。

【例4.21】filter()方法示例。

```
<div id="app">
    <ul>
        <li v-for="n in items">{{ n }}</li>
    </ul>
</div>
<script>
    var app=new Vue({
        el:"#app",
        data:{
            numbers: [ 1, 2, 3, 4, 5 ]
        },
        computed:{
            items: function () {
                return this.numbers.filter(function (number)
                {
                    return number<4
                })
            }
        }
    })
</script>
```

在IE 11浏览器里面运行的结果如图4-25所示。

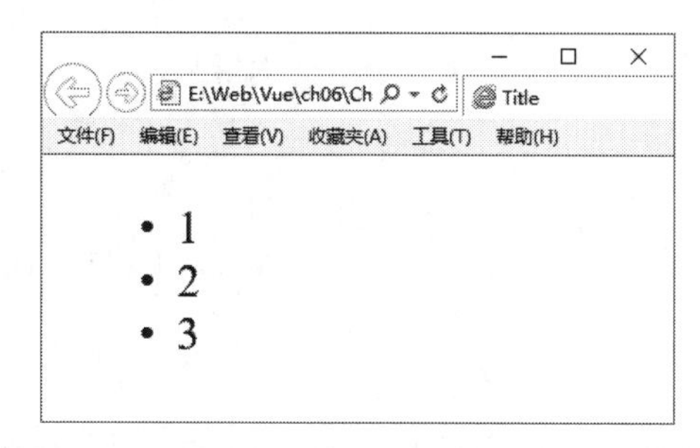

图4-25　filter()方法示例

读者可能会认为，这将导致Vue丢弃现有DOM并重新渲染整个列表，事实并非如此，Vue为了使得DOM元素得到最大范围的重用而实现了一些智能的启发式方法，所以用一个含有相同元素的数组去替换原来的数组是非常高效的操作。

4.3.4 对象变更检测注意事项

由于JavaScript的限制，Vue不能检测对象属性的添加或删除。

Vue实现数据双向绑定的过程：当把一个普通的JavaScript对象传给Vue实例的data选项，Vue将遍历此对象所有的属性，并使用Object.defineProperty()把这些属性全部转为getter/setter。每个组件实例都有相应的watcher实例对象，它会在组件渲染的过程中把属性记录为依赖，之后当依赖项的setter被调用时，会通知watcher重新计算，从而使它关联的组件得以更新，实现数据data变化，更新视图view。

若一个对象的属性没有在data中声明，则它就不是响应式的。Vue会在初始化实例时对属性执行getter/setter转化过程，因此这个对象属性就是响应式的。而执行这个过程必须在data中声明才会有。

例如下面代码：

```
var app = new Vue({
  data:{
    a:1
  }
})
app.b=2;
```

其中app.a是响应式的，app.b不是响应式的。

对于已经创建的实例，Vue不允许动态添加根级别的响应式属性。但是，可以使用Vue.set(object, propertyName, value)方法向嵌套对象添加响应式属性。例如：

```
var app = new Vue({
  data: {
    object: {
      name: '小明'
    }
  }
})
```

可以添加一个新的age属性到嵌套的object对象：

```
Vue.set(app.object, 'age', 27)
```

还可以使用app.$set实例方法，它只是全局Vue.set的别名：

```
app.$set(app.object, 'age', 27)
```

有时可能需要为已有对象赋值多个新属性，例如使用Object.assign()或_.extend()。

在这种情况下，应该用两个对象的属性创建一个新的对象。所以，要想添加新的响应式属性，不要像这样编写：

```
Object.assign(app.object, {
  age: 27,
  sex: '男'
})
```

应该这样编写：

```
app.userProfile = Object.assign({},app.object,{
  age: 27,
  sex: '男'
})
```

4.3.5 在 <template> 上使用 v-for

类似于 v-if，也可以利用带有 v-for 的 <template> 来循环渲染一段包含多个元素的内容。

【例 4.22】在 <template> 上使用 v-for。

```
<div id="app">
    <template v-for="n in items">
         <li>{{n.name}}-{{n.age}}-{{n.sex}}</li>
        <hr/>
    </template>
</div>
<script>
    var app = new Vue({
        el:'#app',
        data:{
            items:[
                {
                    name:'小明',
                    age:15,
                    sex:'男'
                },
                {
                    name:'小红',
                    age:15,
                    sex:'女'
                }
            ]
        }
    })
</script>
```

在 IE 11 浏览器里面运行的结果如图 4-26 所示。

图 4-26　DOM 渲染结果

提示

template 中可以放执行语句，最终编译后不会被渲染成元素。一般常和 v-for 及 v-if 一起结合使用，这样会使得整个 HTML 结构没有那么多多余的元素，整个结构会更加清晰。

4.3.6　v-for 与 v-if 一同使用

当它们处于同一节点上时，v-for 的优先级比 v-if 更高，这意味着 v-if 将分别重复运行于每个 v-for 循环中。当只想为部分项渲染节点时，这种优先级的机制会十分有用。例如下面示例，循环出没有报到的学生名字。

【例 4.23】v-for 与 v-if 一同使用。

```
<div id="app">
    <h3>没有报到的学生名单：</h3>
    <ul>
            <li v-for="n in items" v-if="!n.value">
                {{ n.name}}
            </li>
        </ul>
</div>
<script>
    var app = new Vue({
        el:'#app',
        data:{
            items:[
                {name:'小明',},
                {name:'小红', value:'已报到'},
                {name:'小华', value:'已报到'},
                {name:'小思'}
            ]
        }
```

```
    })
</script>
```

在 IE 11 浏览器里面运行的结果如图 4-27 所示。

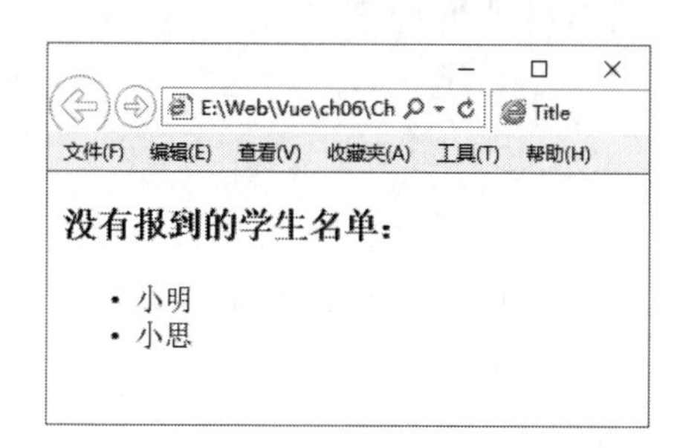

图 4-27　v-for 与 v-if 一起作用效果

注意

不推荐在同一元素上使用 v-if 和 v-for，必要情况下应该替换成计算属性 computed。

【例 4.24】替换成计算属性。

```
<div id="app">
    <h3>没有报到的学生名单：</h3>
    <ul>
        <li v-for="n in student">
            {{ n.name}}
        </li>
    </ul>
</div>
<script>
    var app = new Vue({
        el:'#app',
        data:{
            items:[
                {name:'小明',},
                {name:'小红',},
                {name:'小华', value:'已报到'},
                {name:'小思'}
            ]
        },
        computed:{
            student:function(){
                return this.items.filter(function (n) {
                    return !n.value
                })
            }
        }
    })
</script>
```

在 IE 11 浏览器里面运行的结果如图 4-28 所示。

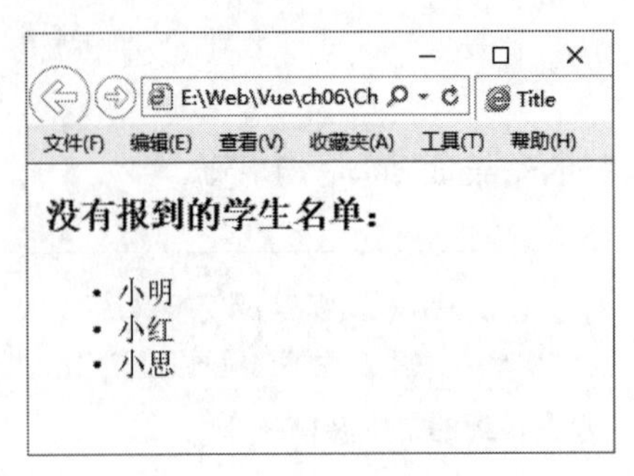

图 4-28　计算属性实现效果

4.4 自定义指令

前面小节介绍过了许多 Vue 内置的指令，例如 v-if、v-show 等，这些丰富的内置指令能满足绝大部分业务需求，不过在需要一些特殊功能或者希望对 DOM 进行底层的操作时，就要用到自定义指令。

自定义指令的注册方法和组件很像，也分全局注册和局部注册，例如注册一个 v-focus 的指令，用于在＜ input>、＜ textarea> 元素初始化时自动获得焦点，两种写法分别是：

```
//全局注册
Vue.directive('focus',{
    //指令选项
});
//局部注册
var app=new Vue({
    el:'#app',
    directives:{
        focus:{
            //指令选项
        }
    }
})
```

写法与组件基本类似，只是方法名由 component 改为了 directive。上面示例只是注册了自定义指令 v-focus，还没有实现具体功能，下面具体介绍自定义指令的各个选项。

- bind：只调用一次，指令第一次绑定到元素时调用，用这个钩子函数可以定义一个在绑定时执行一次的初始化动作。
- update：被绑定元素所在的模板更新时调用，而不论绑定值是否变化。通过比较更新前后的绑定值，可以忽略不必要的模板更新。
- inserted：被绑定元素插入父节点时调用（父节点存在即可调用，不必存在于 document 中）。
- componentUpdated：被绑定元素所在模板完成一次更新周期时调用。
- unbind：只调用一次，指令与元素解绑时调用。

可以根据需求在不同的钩子函数内完成逻辑代码，例如，上面的 v-focus，我们希望在元素插入父节点时就调用，那用到的最好是 inserted 选项。

【例 4.25】自定义 v-focus 指令。

```
<div id="app">
    <input v-focus>
</div>
<script>
    //注册一个全局自定义指令 v-focus
    Vue.directive('focus', {
        //当被绑定的元素插入到DOM中时
        inserted: function (el) {
            //聚焦元素
            el.focus()
        }
    });
    new Vue({
        el: '#app'
    });
</script>
```

在 IE 11 浏览器里面运行，打开页面，input 输入框就自动获得了焦点，成为可输入状态，效果如图 4-29 所示。

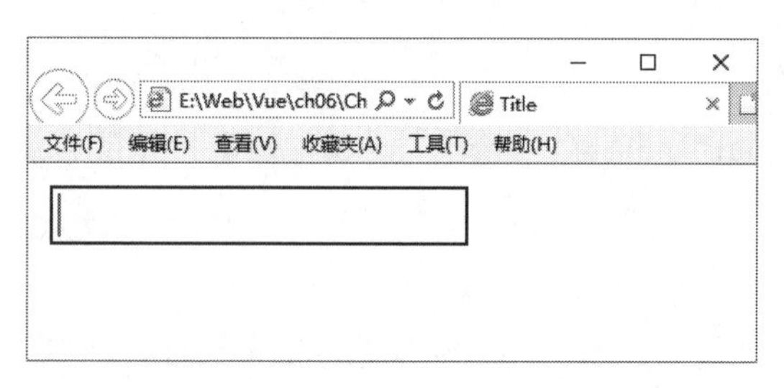

图 4-29　自定义 v-focus 指令的效果

每个钩子函数都有几个参数可用，例如上面用到的 el。它们的含义如下。

- el：指令所绑定的元素，可以用来直接操作 DOM。
- binding：一个对象，包含以下属性。
  - name：指令名，不包括 v- 前缀。
  - value：指令的绑定值，例如 v-my-directive = "1 +1"，value 的值是 2。
  - oldValue：指令绑定的前一个值，仅在 update 和 componentUpdated 钩子中可用。无论值是否改变都可用。
  - expression：绑定值的字符串形式。例如 v-my-directive="1+1"， expression 的值是 "1 +1"。
  - arg：传给指令的参数。例如 v-my-directive：foo，arg 的值是 foo。
  - modifiers：一个包含修饰符的对象。例如 v-my-directive.foo.bar，修饰符对象 modifiers 的值是 { foo: true,bar:true } 。
- vnode：Vue 编译生成的虚拟节点。
- oldVnode：上一个虚拟节点，仅在 update 和 componentUpdated 钩子中可用。

注意

除了 el 之外，其他参数都应该是只读的，切勿进行修改。如果需要在钩子之间共享数据，建议通过元素的 dataset 来进行。

下面是结合了以上参数的一个具体示例。

【例 4.26】钩子函数的参数示例。

```
<div id="app">
    <div id="hook-arguments-example" v-demo:foo.a.b="message"></div>
</div>
<script>
    Vue.directive('demo', {
        bind: function (el, binding, vnode) {
            var s = JSON.stringify
            el.innerHTML =
                'name: '       + s(binding.name) + '<br>' +
                'value: '      + s(binding.value) + '<br>' +
                'expression: ' + s(binding.expression) + '<br>' +
                'argument: '   + s(binding.arg) + '<br>' +
                'modifiers: '  + s(binding.modifiers) + '<br>' +
                'vnode keys: ' + Object.keys(vnode).join(', ')
        }
    })
    new Vue({
        el: '#app',
        data: {
            message: 'hello!'
        }
    })
</script>
```

在 IE 11 浏览器里面运行，div 的内容会使用 innerHTML 重置，结果如图 4-30 所示。

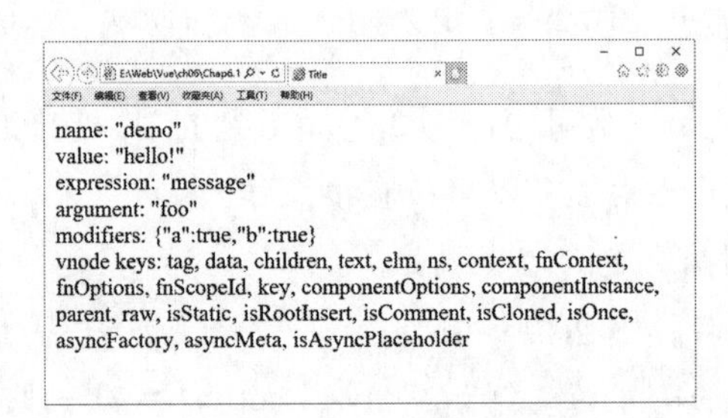

图 4-30　div 的内容

4.5 疑难解惑

疑问 1：Vue 2.0 中，v-for 中的 :key 到底有什么用？

当 Vue 用 v-for 更新已渲染过的元素列表时，它默认用“就地复用”策略。如果数据项的顺序被改变，Vue 将不会移动 DOM 元素来匹配数据项的顺序，而是简单复用此处每个元素，并且确保它在特定索引下显示已被渲染过的每个元素。

这个默认的模式是高效的，但是只适用于不依赖子组件状态或临时 DOM 状态（例如：表单输入值）的列表渲染输出。

为了给 Vue 一个提示，以便它能跟踪每个节点的身份，从而重用和重新排序现有元素，需要为每项提供一个唯一 key 属性。理想的 key 值是每项都有的且唯一的 id。但它的工作方式类似于一个属性，所以需要用 v-bind 来绑定动态值：

```
<div v-for="item in items" :key="item.id">
  <!-- 内容 -->
</div>
```

建议尽可能在使用 v-for 时提供 key，除非遍历输出的 DOM 内容非常简单，或者是刻意依赖默认行为以获取性能上的提升。

疑问 2：由于 JavaScript 的限制，Vue 不能检测数组的变动，当利用索引直接设置一个数组项时，例如：vm.items[index]=newValue，可以使用什么方法来实现？可以使用 Vue.set() 来实现，例如下面示例：

```
<div id="app">
    <li v-for="n in items">{{n}}</li>
</div>
<script>
    var vm = new Vue({
        el:'#app',
        data: {
            items: ['a', 'b', 'c']
        }
    })
    Vue.set(vm.items,3, "d")
    Vue.set(vm.items,5, "e")
</script>
```

在谷歌浏览器中运行的结果如图 4-31 所示。

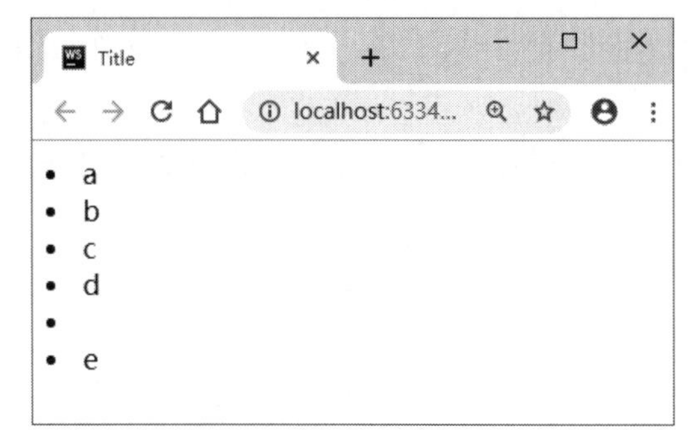

图 4-31　数组遍历效果

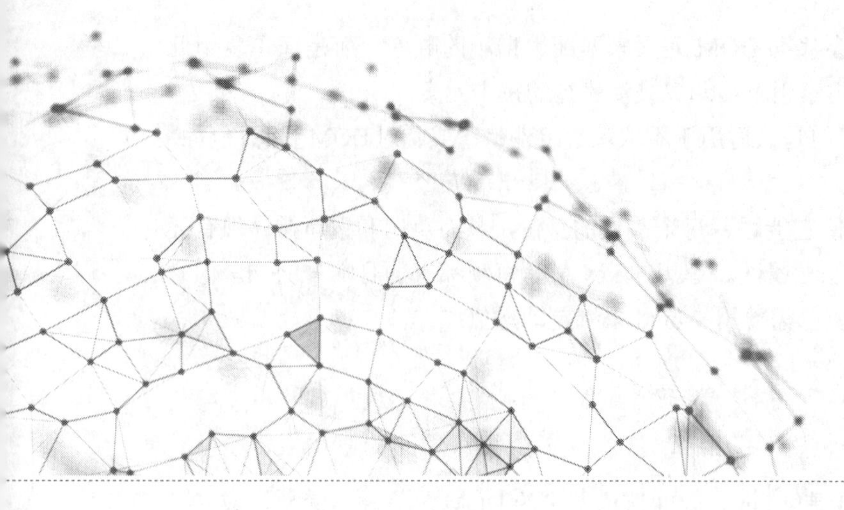

第5章

页面元素样式的绑定

在 Vue 中，操作元素的 class 列表和内联样式是数据绑定的一个常见需求。因为它们都是属性，所以可以用 v-bind 处理它们：只需要通过表达式计算出字符串结果即可。不过，字符串拼接麻烦且易错，因此，在将 v-bind 用于 class 和 style 时，Vue 做了专门的增强。表达式结果的类型除了字符串之外，还可以是对象或数组。

5.1 绑定 HTML 样式

在 Vue 中，动态的样式类在 v-on:class 中定义，静态的类名写在 class 样式中。

5.1.1 数组语法

Vue 中提供了使用数组进行绑定样式的方式，可以直接在数组中写上样式的类名。

注意

如果不使用单引号包裹类名，它其实代表的还是一个变量的名称，会出现错误信息。

【例 5.1】Class 数组语法。

```
<style>
        .static{
            color: white;             /*定义字体颜色*/
        }
        .style1{
            background: #4f43ff;      /*定义背景颜色*/
```

```
        }
        .style2{
            width: 200px;           /*定义宽度*/
            height: 100px;           /*定义高度*/
        }
    </style>
<div id="app">
    <div class="static" v-bind:class="['style1','style2']">{{message}}</div>
</div>
<script>
    new Vue({
        el: '#app',
        data:{
            message:"数组语法"
        }
    })
</script>
```

在 IE 11 浏览器里面运行，打开控制台，可以看到 DOM 渲染的样式，如图 5-1 所示。

图 5-1　数组语法渲染结果

如果想以变量的方式，就需要先定义好这个变量。示例中的样式与上例样式相同。

```
<div id="app">
    <div class="static" v-bind:class="[Class1,Class2]">{{message}}</div>
</div>
<script>
    new Vue({
        el: '#app',
        data: {
            message:'数组语法',
            Class1:'style1',
            Class2:'style2'
        }
    })
</script>
```

在数组语法中也可以使用对象语法，来控制样式是否使用。下面示例中的样式与上例样式相同。

```
<div id="app">
    <div class="static" v-bind:class="[{style1:boole}, 'style2']">{{message}}</
div>
</div>
<script>
    new Vue({
        el: '#app',
        data: {
            message:'数组语法',
            boole:true
        }
    })
</script>
```

在 IE 11 浏览器中运行，渲染的结果和上面示例相同。

5.1.2 对象语法

在上面小节的最后，在数组中使用了对象来设置样式，在 Vue 中也可以直接使用对象来设置样式。对象的属性为样式的类名，value 则为 true 或者 false，当值为 true 时显示样式。由于对象的属性可以带引号，也可不带引号，所以属性就按照自己的习惯写法就可以了。

【例 5.2】Class 对象语法。

```
<style>
    .static{
        color: white;              /*定义字体颜色*/
    }
    .style1{
        background: #4f43ff;       /*定义背景颜色*/
    }
    .style2{
        width: 200px;              /*定义宽度*/
        height: 100px;             /*定义高度*/
    }
</style>
<div id="app">
    <div class="static" v-bind:class="{ style1: boole1, 'style2':
      boole2}">{{message}}</div>
</div>
<script>
    new Vue({
        el: '#app',
        data: {
            boole1: true,
            boole2: true,
            message:"对象语法"
        }
    })
</script>
```

在 IE 11 浏览器里面运行，打开控制台，可以看到 DOM 渲染的样式，如图 5-2 所示。

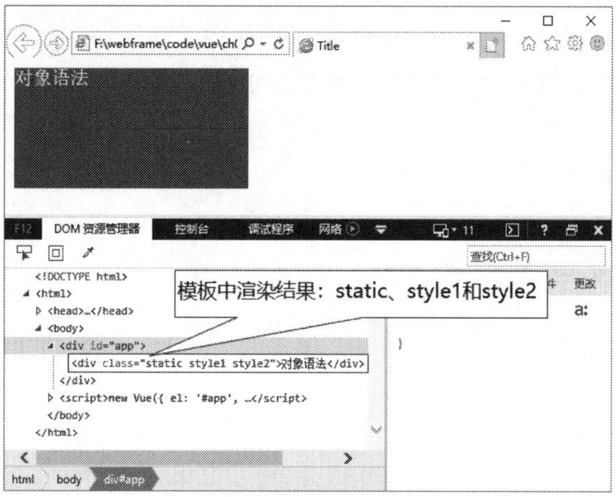

图 5-2　Class 对象语法

当 style1 或 style2 变化时，class 列表将相应地更新。例如，如果 style2 的值变更为 false。

```
<script>
    new Vue({
        el: '#app',
        data: {
            boole1: true,
            boole2: false,
            message:"对象语法"
        }
    })
</script>
```

在 IE 11 浏览器里面运行，打开控制台，可以看到 DOM 渲染的样式，如图 5-3 所示。

图 5-3　模板渲染结果

当对象中的属性过多，如果还是全部写到元素上时，就会显得比较繁多。这时可以在元素上只写上对象变量，在 Vue 实例中进行定义。下面示例中样式和上面示例样式相同。

```
<div id="app">
    <div class="static" v-bind:class="objStyle">{{message}}</div>
</div>
<script>
    new Vue({
        el: '#app',
        data: {
            message:"对象语法",
            objStyle:{
                style1: true,
                style2: true
            }
        }
    })
</script>
```

在 IE 11 浏览器中运行，渲染的结果和上面示例相同。

也可以绑定一个返回对象的计算属性，这是一个常用且强大的模式。下面示例中样式和上面样式相同。

```
<div id="app">
    <div class="static" v-bind:class="classObject">{{message}}</div>
</div>
<script>
    new Vue({
        el: '#app',
        data: {
            message:'对象语法',
            boole1: true,
            boole2: true
        },
        computed: {
            classObject: function () {
                return {
                    style1:this.boole1,
                    'style2':this.boole2
                }
            }
        }
    })
</script>
```

在 IE 11 浏览器中运行，渲染的结果和上面示例相同。

5.1.3 在自定义组件上使用 class

当在一个自定义组件上使用 class 属性时，这些类将被添加到该组件的根元素上面。这个元素上已经存在的类不会被覆盖。

例如，声明组件 my-component 如下：

```
Vue.component('my-component', {
  template: '<p class="style1 style2">Hello</p>'
})
```

然后在使用它的时候添加一些 class 样式，例如 style3 和 style4：

```
<my-component class=" style3 style4"></my-component>
```

HTML 将被渲染为：

```
<p class=" style1 style2 style3 style4">Hello</p>
```

对于带数据绑定的 class 也同样适用：

```
<my-component v-bind:class="{ style5: isActive }"></my-component>
```

当 isActive 为 Truthy 时，HTML 将被渲染为：

```
<p class=" style1 style2 style5">Hello</p>
```

提示

在 JavaScript 中，Truthy(真值)指的是在布尔值上下文中转换后的值为真的值。所有值都是真值，除非它们被定义为 falsy(即除了 false，0，""，null、undefined 和 NaN 外)。

5.2 绑定内联样式

内联样式是将 CSS 样式编写到元素的 style 属性中。

5.2.1 对象语法

与使用属性为元素设置 class 样式相同，在 Vue 中，也可以使用对象为元素设置 style 样式。

v-bind:style 的对象语法十分直观——看着非常像 CSS，但其实它是一个 JavaScript 对象。CSS 属性名可以用驼峰式（camelCase）或短横线分隔（kebab-case，用引号包裹起来）来命名。

【例 5.3】style 对象语法示例。

```
<div id="app">
    <div v-bind:style="{color:'red',fontSize:'30'}">对象语法</div>
</div>
<script>
    new Vue({
        el: '#app',
    })
</script>
```

在 IE 11 浏览器里面运行，打开控制台，渲染结果如图 5-4 所示。

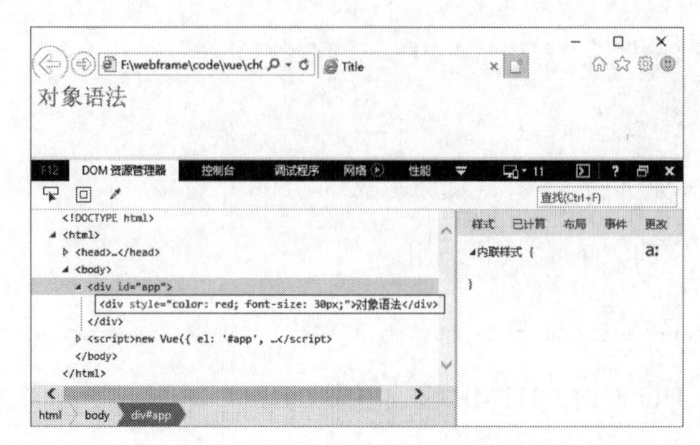

图 5-4　style 对象语法示例

也可以在 Vue 实例对象中定义属性，用来代替样式值，例如下面代码：

```
<div id="app">
    <div v-bind:style="{ color: styleColor, fontSize: fontSize + 'px' }">对象语
      法</div>
</div>
<script>
    new Vue({
        el: '#app',
        data: {
            styleColor: 'red',
            fontSize: 30
        }
    })
</script>
```

在 IE 11 浏览器中运行效果和上例相同。

同样，可以直接绑定一个样式对象变量，这样代码看起来也会更简洁美观。

```
<div id="app">
    <div v-bind:style="styleObject">对象语法</div>
</div>
<script>
    new Vue({
        el: '#app',
        data: {
            styleObject: {
                color: 'blue',
                fontSize: '30px'
            }
        }
    })
</script>
```

在 IE 11 浏览器里面运行，打开控制台，渲染结果如图 5-5 所示。

图 5-5　渲染结果

同样，对象语法常常结合返回对象的计算属性使用。

```
<div id="app">
    <div v-bind:style="styleObject">对象语法</div>
</div>
<script>
    new Vue({
        el: '#app',
        //计算属性
        computed:{
            styleObject:function(){
                return {
                    color: 'blue',
                    fontSize: '30px'
                }
            }
        }
    })
</script>
```

在 IE 11 浏览器中运行，渲染的结果和上面示例相同。

5.2.2　数组语法

v-bind:style 的数组语法可以将多个样式对象应用到同一个元素上，样式对象可以是 data 中定义的样式对象和计算属性中 return 的对象。

【例 5.4】style 数组语法示例。

```
<div id="app">
    <div v-bind:style="[styleObject1,styleObject2]">数组语法</div>
</div>
<script>
    new Vue({
        el: '#app',
        data: {
            styleObject1: {
                color: 'blue',
                fontSize: '30px'
            }
```

```
            },
            //计算属性
            computed:{
                styleObject2:function(){
                    return {
                        border: '1px solid red',
                        padding: '30px'
                    }
                }
            }
        })
</script>
```

在 IE 11 浏览器里面运行，打开控制台，渲染结果如图 5-6 所示。

图 5-6　style 数组语法渲染结果

注意

当 v-bind:style 使用需要添加浏览器引擎前缀的 CSS 属性时，例如 transform，Vue 会自动侦测并添加相应的前缀。

5.3 疑难解惑

疑问 1：如果想根据条件切换列表中的 class，如何实现？

可以使用三元表达式。例如下面示例：

```
<style>
    .color1{
        color: white;
    }
    .color2{
```

```
            background: blue;
        }
    </style>
    <div id="app">
        <div v-bind:class="[isActive ? activeClass : '', errorClass]">Hello Vue.
        js</div>
    </div>
    <script>
        var vm = new Vue({
            el:'#app',
            data:{
                isActive:true,
                activeClass:'color1',
                errorClass:'color2'
            }
        })
    </script>
```

上面这种写法，将始终添加 errorClass，只有在 isActive 是 truthy 时才添加 activeClass。

不过，当有多个条件 class 时，这样写有些烦琐，所以在数组语法中也可以使用对象语法：

```
<div v-bind:class="[{ active: isActive }, errorClass]"></div>
```

关于数组语法中也可以使用对象语法的问题，请参考“5.1.1 数组语法”中的示例。

疑问 2：绑定内联样式中的多重值是什么？

从 Vue 2.3.0 起，可以为 style 绑定中的属性提供一个包含多个值的数组，常用于提供多个带前缀的值，例如：

```
<div :style="{ display: ['-webkit-box', '-ms-flexbox', 'flex'] }"></div>
```

这样写只会渲染数组中最后一个被浏览器支持的值。如果浏览器支持不带浏览器前缀的 flexbox，那么就只会渲染 display: flex。

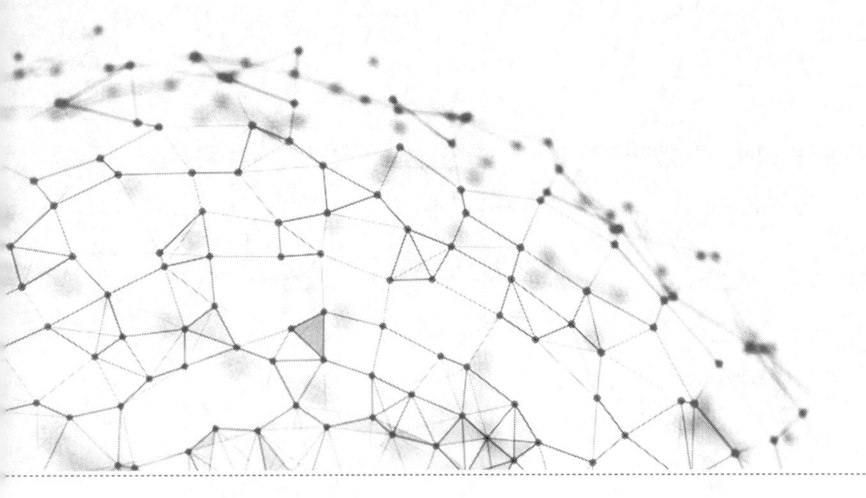

第6章 事件处理

在前面内置指令一章中，简单地介绍了 v-on 的基本用法。本章将详细介绍 Vue 实现绑定事件的方法，使用 v-on 指令监听 DOM 事件来触发一些 JavaScript 代码。

6.1 监听事件

事件其实就是在程序运行当中，可以调用方法，去改变对应的内容。下面先来看一个简单的示例。

```
<div id="app">
    <p>老王的年龄:{{ age }}岁</p>
</div>
<script>
    new Vue({
        el:'#app',
        data:{
            age:"50"
        }
    })
</script>
```

运行的结果为“老王的年龄 :50 岁”。

在上面的示例中，如果想要改变老王的年龄，就可以通过事件来完成。

在 JavaScript 中可以使用的事件，在 Vue 中也都可以使用。使用事件时，需要 v-on 指令监听 DOM 事件。

下面我们在上面示例中添加两个按钮，当单击按钮时增加或减少老王的年龄。

【例 6.1】添加单击事件。

```
<div id="app">
    <button v-on:click="age--">减少1岁</button>
    <button v-on:click="age++">增加1岁</button>
    <p>老王的年龄:{{ age }}岁</p>
</div>
<script>
    new Vue({
        el:'#app',
        data:{
            age:"50"
        }
    })
</script>
```

在IE 11浏览器里面运行，不断单击“增加1岁”按钮，老王的年龄不断增长，如图6-1所示。

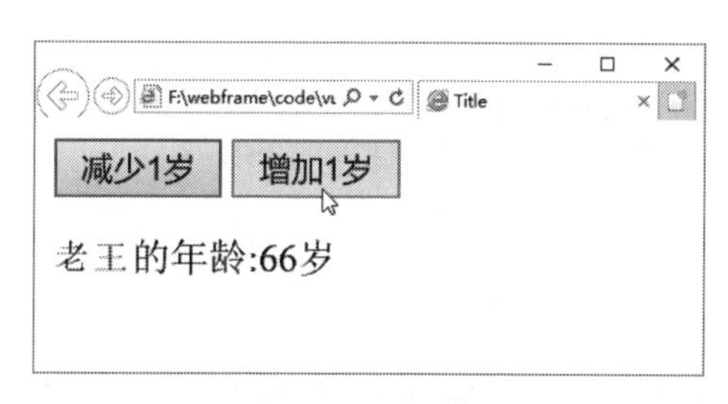

图 6-1　单击事件

为什么在HTML中监听事件呢？这种事件监听的方式违背了关注点分离这个长期以来的优良传统。但不必担心，因为所有的Vue事件处理方法和表达式都严格绑定在当前视图的ViewModel上，它不会导致任何维护上的困难。实际上，使用v-on有3个好处。

（1）扫一眼HTML模板，便能轻松定位在JavaScript代码里对应的方法。

（2）因为无须在JavaScript里手动绑定事件， ViewModel代码可以是非常纯粹的逻辑，和DOM完全解耦，更易于测试。

（3）当一个ViewModel被销毁时，所有的事件处理器都会自动被删除，无须担心如何清理它们。

6.2 事件处理方法

在上一节介绍的示例中，我们是直接操作属性，但在实际的项目开发中，是不可能直接对属性进行操作的。例如，在上面的案例中，如果想要单击一次按钮，使老王的年龄增加或减少10岁呢？

许多事件处理逻辑会更为复杂，所以直接把JavaScript代码写在v-on指令中是不可行的。在Vue中，v-on还可以接收一个需要调用的方法名称，我们可以在方法中来完成复杂的逻辑。

下面我们在方法中来实现单击按钮增加或减少 10 岁的操作。

【例 6.2】事件处理方法。

```
<div id="app">
    <button v-on:click="reduce">减少10岁</button>
    <button v-on:click="add">增加10岁</button>
    <p>老王的年龄:{{ age }}岁</p>
</div>
<script>
    new Vue({
        el:'#app',
        data:{
            age:50
        },
        methods:{
            add:function(){
                this.age+=10
            },
            reduce:function(){
                this.age-=10
            }
        }
    })
</script>
```

在 IE 11 浏览器里面运行，单击“减少 10 岁”按钮，老王的年龄减少 10 岁，如图 6-2 所示。

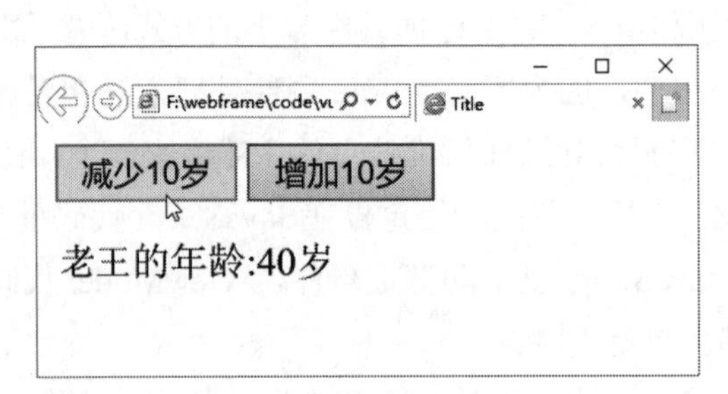

图 6-2　事件处理方法

提示

“v-on:”可以使用“@”代替，例如下面代码：

```
<button @click="reduce">减少10岁</button>
<button @click="add">增加10岁</button>
```

v-on: 和 @ 作用是一样的，可以根据自己的喜好进行选择。

这样就把逻辑代码写到了方法中。相对于上面示例，还可以通过传入参数来实现，在调用方法时，传入想要增加或减少的数量，在 Vue 中定义一个 change 参数来接收。

【例 6.3】事件处理方法的参数。

```
<div id="app">
    <button v-on:click="reduce(10)">减少10岁</button>
```

```
        <button v-on:click="add(10)">增加10岁</button>
        <p>老王的年龄:{{ age }}岁</p>
    </div>
    <script>
        new Vue({
            el:'#app',
            data:{
                age:50
            },
            methods:{
               //在方法中定义一个参数change，接收HTML中传入的参数
                add:function(change){
                    this.age+=change
                },
                reduce:function(change){
                    this.age-=change
                }
            }
        })
    </script>
```

在 IE 11 浏览器里面运行，单击“增加 10 岁”按钮，老王的年龄增加 10 岁，如图 6-3 所示。

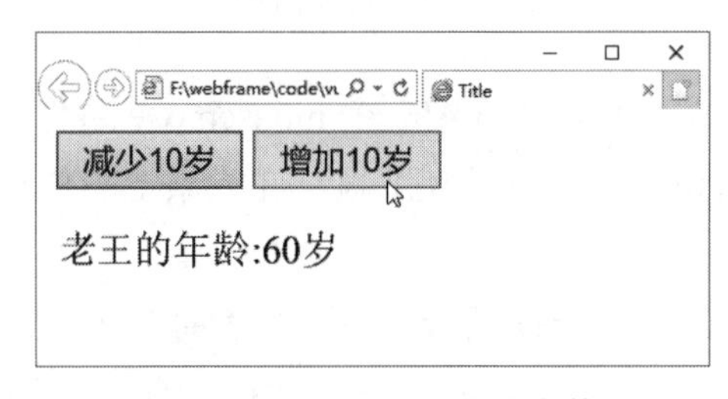

图 6-3　事件处理方法的参数

对于定义的方法，多个事件都可以调用。例如，在上面的示例中，再添加两个按钮，分别添加双击事件，并调用 add() 和 reduce() 方法。单击事件传入参数 1，双击事件传入参数 10，在 Vue 中使用 change 进行接收。

【例 6.4】多个事件调用方法。

```
    <div id="app">
        <div>单击:
            <button v-on:click="reduce(1)">减少1岁</button>
            <button v-on:click="add(1)">增加1岁</button>
        </div>
        <p>老王的年龄:{{ age }}岁</p>
        <div>双击:
            <button v-on:dblclick="reduce(10)">减少10岁</button>
            <button v-on:dblclick="add(10)">增加10岁</button>
        </div>
    </div>
    <script>
        new Vue({
            el:'#app',
            data:{
                age:50
```

```
            },
            methods:{
                add:function(change){
                    this.age+=change
                },
                reduce:function(change){
                    this.age-=change
                }
            }
        })
</script>
```

在IE 11浏览器中运行，单击或者双击按钮，老王的年龄随着改变，效果如图6-4所示。

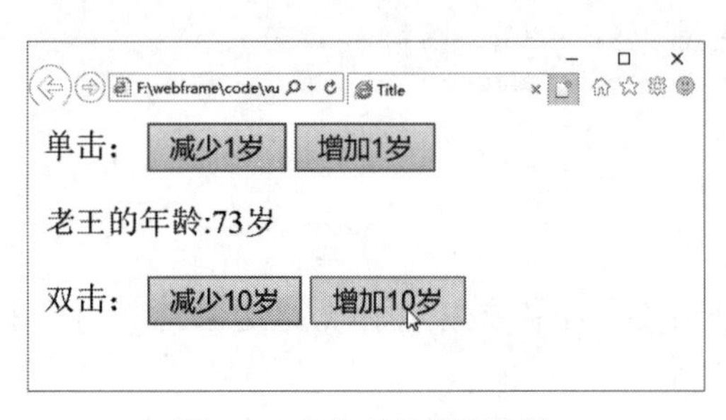

图6-4　多个事件调用方法

除了上述的单击和双击事件，再介绍一个mousemove事件，当鼠标在元素内部移动时会不断触发该事件，可以通过该事件获取鼠标在元素上的位置。下面通过一个示例进行介绍。

在示例中首先定义一个area元素，并设置简单样式，然后使用v-on:绑定mousemove事件。在Vue中定义position方法，通过event事件对象获取鼠标的位置，并赋值给x和y，最后在页面中渲染x和y。

提示

event对象代表事件的状态，例如事件在其中发生的元素、键盘按键的状态、鼠标的位置、鼠标按键的状态等。当一个事件发生的时候，和当前这个对象发生的这个事件有关的一些详细信息都会被临时保存到一个指定的地方——event对象，供我们在需要的时候调用。这个对象是在执行事件时，浏览器通过函数传递过来的。

【例6.5】mousemove事件。

```
<style>
        .area{
            width: 400px;               /*定义宽度*/
            height: 200px;              /*定义高度*/
            border:1px solid black;     /*定义边框*/
            text-align:center;          /*水平居中*/
            line-height:200px;          /*定义行高*/
            font-size: 20px;            /*定义字体大小*/
        }
```

```
    </style>
<div id="app">
    <div class="area" @mousemove="position">
        {{x}},{{y}}
    </div>
</div>
<script>
    new Vue({
        el:'#app',
        data:{
            x:0,
            y:0
        },
        methods:{
            position:function(event){
                console.log(event)
                //获取鼠标的坐标点
                this.x=event.offsetX;
                this.y=event.offsetY;
            }
        }
    })
</script>
```

在 IE 11 浏览器里面运行，在 area 元素中移动鼠标，元素中间显示鼠标的相对位置，如图 6-5 所示。

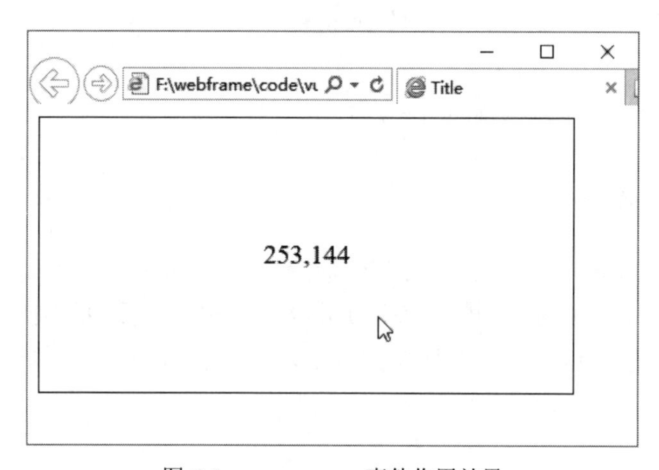

图 6-5　mousemove 事件作用效果

注意

在 Vue 事件中，可以使用事件名称 add 或 reduce 进行调用，也可以使用事件名加上“()”的形式，例如 add()、reduce()。但是在具有参数时需要使用 add()、reduce() 的形式。在 {{}} 中调用方法时，必须使用 add()、reduce() 的形式。

6.3 事件修饰符

对事件添加一些通用的限制，例如阻止事件冒泡，Vue 则对这种事件的限制提供了特定的写法，语法如下：

```
v-on:事件.修饰符
```

在事件处理程序中调用 event.preventDefault()（阻止默认行为）或 event.stopPropagation()（阻止事件冒泡）是非常常见的需求。尽管可以在方法中轻松实现这一点，但更好的方式是使用纯粹的数据逻辑，而不是去处理 DOM 事件细节。

在 Vue 中，事件修饰符处理了许多 DOM 事件的细节，让我们不再需要花大量的时间去处理这些烦恼的事情，而能有更多的精力专注于程序的逻辑处理。在 Vue 中事件修饰符主要有以下几个。

- .stop：等同于 JavaScript 中的 event.stopPropagation()，阻止事件冒泡。
- .capture：与事件冒泡的方向相反，事件捕获由外到内。
- .self：只会触发自己范围内的事件。
- .once：只会触发一次。
- .prevent：等同于 JavaScript 中的 event.preventDefault()，阻止默认事件的的发生
- .passive：执行默认行为。

下面分别来介绍每个修饰符的用法。

6.3.1 stop 修饰符

stop 修饰符用来阻止事件冒泡。在下面的示例中，创建了一个 div 元素，在其内部也创建一个 div 元素，并分别为它们添加单击事件。根据事件的冒泡机制可以得知，当单击内部的 div 元素之后，会扩散到父元素 div，从而触发父元素的单击事件。

【例 6.6】冒泡事件。

```
<style>
        .outside{
            width: 200px;                    /*定义宽度*/
            height: 200px;                   /*定义高度*/
            border: 1px solid red;           /*定义边框*/
            text-align: center;              /*文本水平居中*/
        }
        .inside{
            width: 100px;                    /*定义宽度*/
            height: 100px;                   /*定义高度*/
            border:1px solid black;          /*定义边框*/
            margin:  25%;                    /*定义外边距*/
        }
    </style>
<body>
<div id="app">
```

```
        <div class="outside" @click="outside">
            <div class="inside" @click ="inside">冒泡事件</div>
        </div>
    </div>
    <script>
        new Vue({
            el: '#app',
            methods: {
                outside: function () {
                    alert("外面的div")
                },
                inside: function () {
                    alert("内部的div")
                }
            }
        })
    </script>
```

在 IE 11 浏览器里面运行，单击内部 inside 元素，触发自身事件，效果如图 6-6 所示；根据事件的冒泡机制，也会触发外部的 outside 元素，效果如图 6-7 所示。

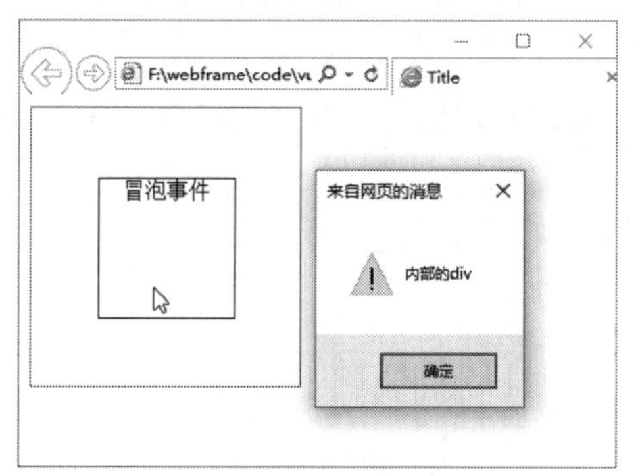

图 6-6　触发内部元素事件

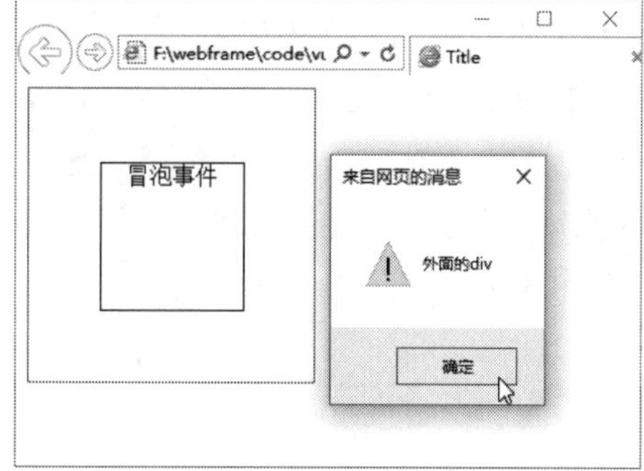

图 6-7　触发外部元素事件

如果不希望出现事件冒泡，则可以使用 Vue 内置的修饰符 stop 便捷地阻止事件冒泡的产生。因为是单击内部 div 元素后产生的事件冒泡，所以只需要在内部 div 元素的单击事件上加上 stop 修饰符即可。

【例 6.7】 stop 修饰符示例。

更改上面 HTML 对应的代码：

```
<div id="app">
    <div class="outside" @click="outside">
        <div class="inside" @click.stop="inside">阻止事件冒泡</div>
    </div>
</div>
```

在 IE 11 浏览器里面运行，单击内部的 inside 后，将不再触发父元素单击事件，如图 6-8 所示。

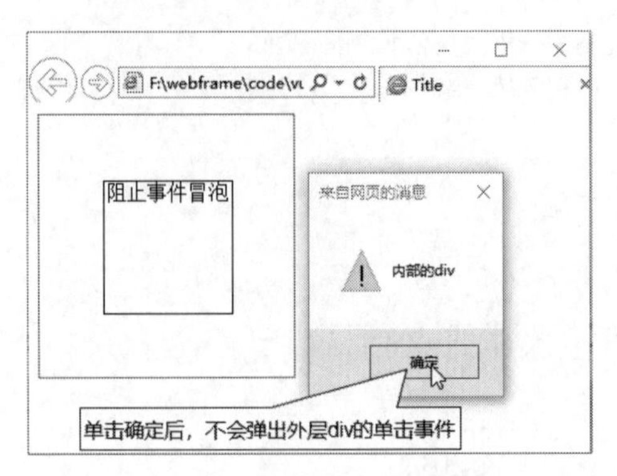

图 6-8　只触发内部元素事件

6.3.2　capture 修饰符

事件捕获模式与事件冒泡模式是一对相反的事件处理流程，当想要将页面元素的事件流改为事件捕获模式时，只需要在父级元素的事件上使用 capture 修饰符即可。若有多个该修饰符，则由外而内触发。

在下面示例中，创建了 3 个 div 元素，把它们分别嵌套，并添加单击事件。为外层的两个 div 元素添加 capture 修饰符，当单击内部的 div 元素时，将从外部向内触发含有 capture 修饰符的 div 元素的事件。

【例 6.8】capture 修饰符示例。

```
<style>
        .outside{
            width: 300px;                          /*定义宽度*/
            height: 300px;                         /*定义高度*/
            color:white;                           /*定义字体颜色*/
            font-size: 30px;
            background: red;                       /*定义背景色*/
        }
        .center{
            width: 200px;                          /*定义宽度*/
            height: 200px;                         /*定义高度*/
            background: #17a2b8;                   /*定义背景色*/
        }
        .inside{
            width: 100px;                          /*定义宽度*/
            height: 100px;                         /*定义高度*/
            background: #a9b4ba;                   /*定义背景色*/
        }
    </style>
<div id="app">
    <div class="outside" @click.capture="outside">
        <div class="center" @click.capture="center">
            <div class="inside" @click="inside">内部</div>
            中间
        </div>
        外层
    </div>
</div>
```

```
<script>
    new Vue({
        el: '#app',
        methods: {
            outside: function () {
                alert("外面的div")
            },
            center: function () {
                alert("中间的div")
            },
            inside: function () {
                alert("内部的div")
            }
        }
    })
</script>
```

在 IE 11 浏览器里面运行，单击外面的 div 元素，会先触发添加了 capture 修饰符的外层 div 元素，如图 6-9 所示；然后触发中间的 div 元素，如图 6-10 所示；最后触发单击的内部元素，如图 6-11 所示。

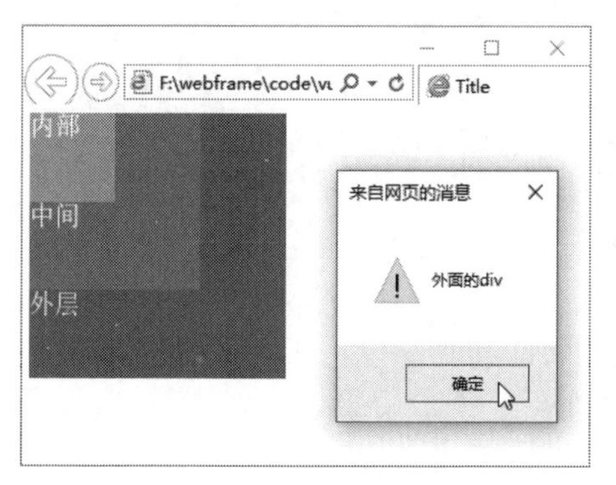

图 6-9　触发外层 div 元素事件

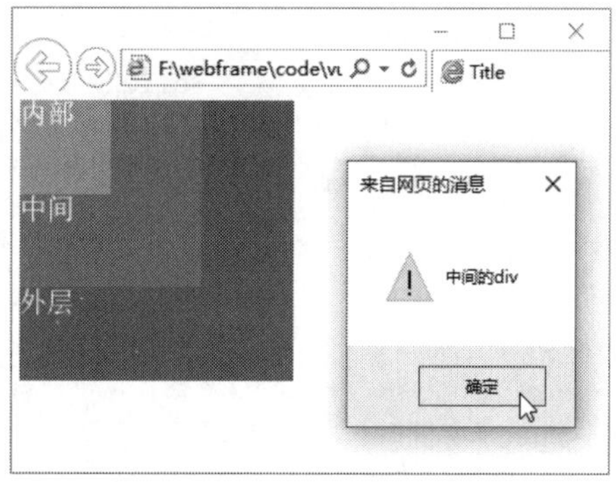

图 6-10　触发中间 div 元素事件

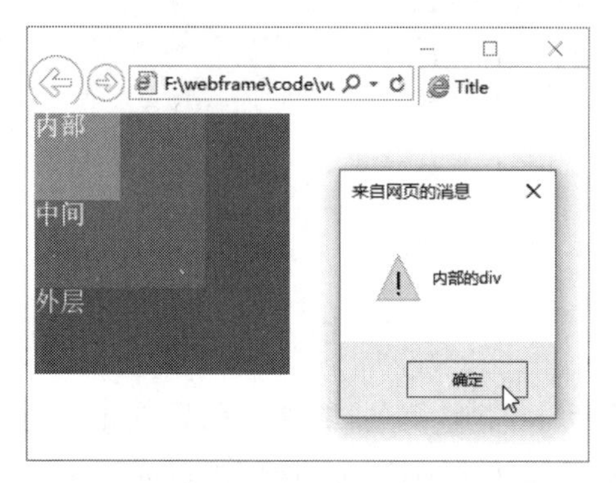

图 6-11　触发内部 div 元素事件

6.3.3　self 修饰符

self 修饰符可以理解为跳过冒泡事件和捕获事件，只有直接作用在该元素上的事件才可以执行。self修饰符会监视事件是否直接作用在元素上，若不是，则冒泡跳过该元素。

【例6.9】self修饰符示例。

```
<style>
    .outside{
        width: 300px;                          /*定义宽度*/
        height: 300px;                         /*定义高度*/
        color:white;                           /*定义字体颜色*/
        font-size: 30px;
        background: red;                       /*定义背景色*/
    }
    .center{
        width: 200px;                          /*定义宽度*/
        height: 200px;                         /*定义高度*/
        background: #17a2b8;                   /*定义背景色*/
    }
    .inside{
        width: 100px;                         /*定义宽度*/
        height: 100px;                        /*定义高度*/
        background: #a9b4ba;                  /*定义背景色*/
    }
</style>
<div id="app">
    <div class="outside" @click="outside">
        <div class="center" @click.self="center">
            <div class="inside" @click="inside">内部</div>
            中间
        </div>
        外层
    </div>
</div>
<script>
    new Vue({
        el: '#app',
        methods: {
            outside: function () {
                alert("外面的div")
            },
            center: function () {
                alert("中间的div")
            },
            inside: function () {
                alert("内部的div")
            }
        }
    })
</script>
```

在IE 11浏览器中运行，单击内部的div后，触发该元素的单击事件，效果如图6-12所示；由于中间div添加了self修饰符，直接单击该元素会跳过；内部div执行完毕，外层的div紧接着执行，效果如图6-13所示。

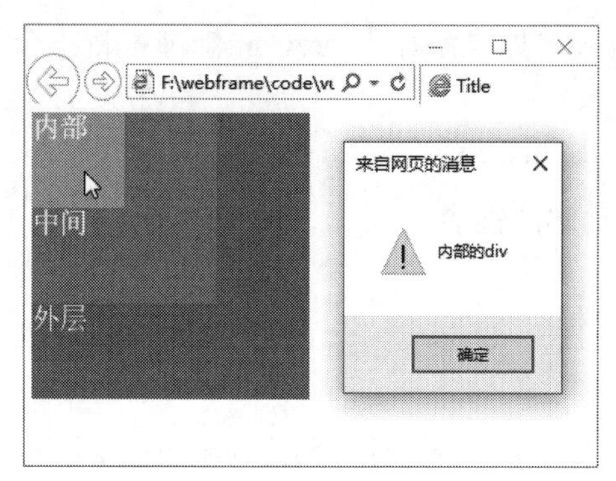

图 6-12　触发内部 div 元素事件

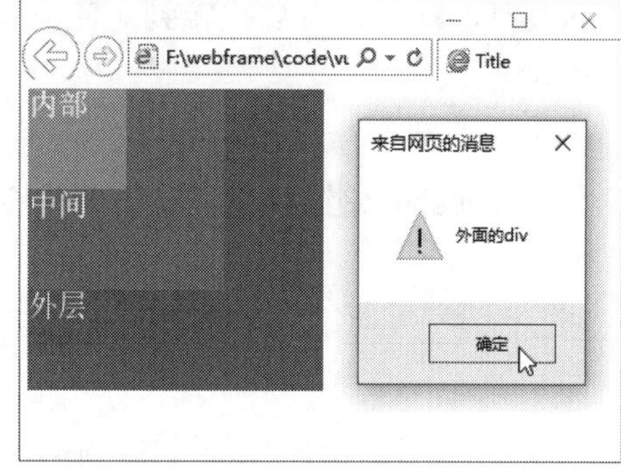

图 6-13　触发外层 div 元素事件

6.3.4　once 修饰符

有时需要只执行一次的操作，例如，微信朋友圈点赞，这时便可以使用 once 修饰符来完成。

提示

不像其他只能对原生的 DOM 事件起作用的修饰符，once 修饰符还能被用到自定义的组件事件上。

【例 6.10】once 修饰符示例。

```
<div id="app">
    <button @click.once="add">点赞 {{ age }}</button>
</div>
<script>
    new Vue({
        el:'#app',
        data:{
            age:0
        },
        methods:{
            add:function(){
                this.age+=1
            },
        }
    })
</script>
```

在 IE 11 浏览器里面运行，单击“点赞 0”按钮，age 值从 0 变成 1，之后，不管再单击多少次，age 的值仍然是 1，效果如图 6-14 所示。

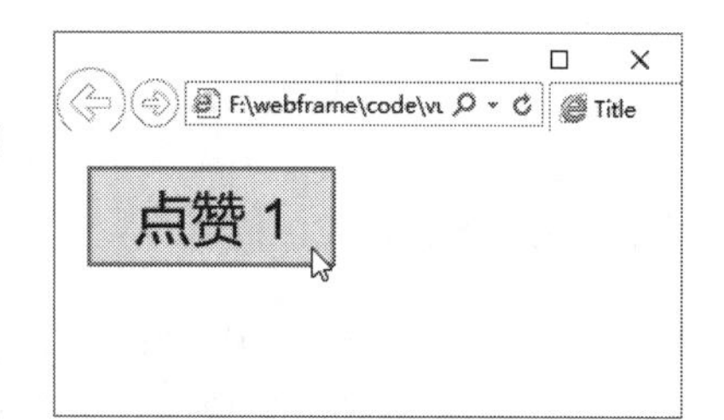

图 6-14　once 修饰符作用效果

6.3.5　prevent 修饰符

prevent 修饰符用于阻止默认行为，例如 <a> 标签，当单击标签时，默认行为会跳转到对应的链接，如果添加上 prevent 修饰符将不会跳转到对应的链接。

提示

passive 修饰符尤其能够提升移动端的性能。

注意

不要把passive修饰符和prevent修饰符一起使用，因为prevent将会被忽略，同时浏览器可能会发生一个警告信息。passive 修饰符会告诉浏览器不想阻止事件的默认行为。

【例 6.11】prevent 修饰符示例。

```
<div id="app">
    <a @click.prevent="alert()" href="https://cn.vuejs.org" >Vue.js官网</a>
</div>
<script>
    new Vue({
        el: '#app',
        methods:{
            alert:function(){
                alert("阻止<a>标签的链接")
            }
        }
    })
</script>
```

在 IE 11 浏览器中运行，单击“Vue.js 官网”，触发 alert() 事件，弹出“阻止 <a> 标签的链接”，效果如图 6-15 所示，单击“确定”按钮，可发现页面将不进行跳转。

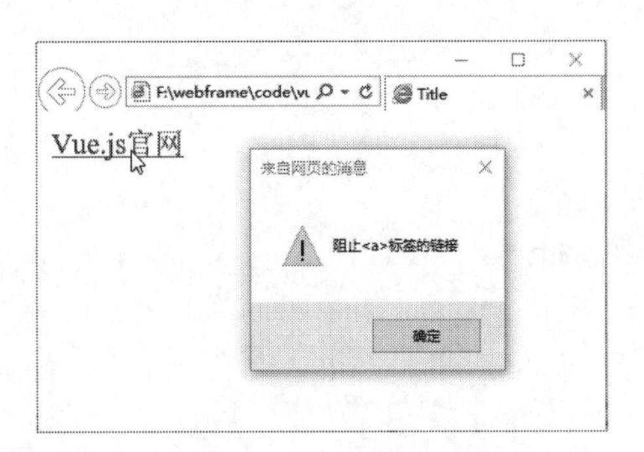

图 6-15　prevent 修饰符

6.3.6　passive 修饰符

明明默认执行的行为，为什么还要使用 passive 修饰符呢？原因是浏览器只有等内核线程执行到事件监听器对应的 JavaScript 代码时，才能知道内部是否会调用 preventDefault 函数，来阻止事件的默认行为，所以浏览器本身是没有办法对这种场景进行优化的。这种场景下，用户的手势事件无法快速产生，会导致页面无法快速执行滑动逻辑，从而让用户感觉到页面卡顿。

通俗说就是每次事件产生，浏览器都会去查询一下是否有 preventDefault 阻止该次事件的默认动作。加上 passive 修饰符就是为了告诉浏览器，不用查询了，没用 preventDefault 阻止默认行为。

passive 修饰符一般用在滚动监听、@scoll 和 @touchmove 中。因为滚动监听过程中，移动每个像素都会产生一次事件，每次都使用内核线程查询 prevent 会使滑动卡顿。通过 passive 修饰符将内核线程查询跳过，可以大大提升滑动的流畅度。

注意

使用修饰符时，顺序很重要。相应的代码会以同样的顺序产生。因此，用 v-on:click.prevent.self 会阻止所有的单击，而 v-on:click.self.prevent 只会阻止对元素自身的单击。

6.4 按键修饰符

在 Vue 中可以使用以下 3 种键盘事件。

- keydown：键盘按键按下时触发。
- keyup：键盘按键抬起时触发。
- keypress：键盘按键按下抬起间隔期间触发。

在日常的页面交互中，经常会遇到这种需求。例如，用户输入账号密码后按 Enter 键，当有一个多筛选条件时，通过单击多选框将自动加载符合选中条件的数据。在传统的前端开发中，当碰到这种类似的需求时，往往需要知道 JavaScript 中需要监听的按键所对应的 keyCode，然后通过判断 keyCode 得知用户是按下了哪个按键，继而执行后续的操作。

提示

keyCode 返回 keypress 事件触发的键值的字符代码或 keydown、keyup 事件的键的代码。

下面来看一个示例，当触发键盘事件时，调用一个方法。在示例中，为两个 input 输入框绑定 keyup 事件，用键盘在输入框输入内容时触发，每次输入内容都会触发并调用 name 或 password 方法。

【例 6.12】触发键盘事件。

```
<div id="app">
    <label for="name">姓名：</label>
    <input v-on:keyup="name" type="text" id="name">
```

```
        <label for="pass">密码: </label>
        <input v-on:keyup="password" type="password" id="pass">
    </div>
    <script>
        new Vue({
            el: '#app',
            methods: {
                name:function(){
                    console.log("正在输入姓名...")
                },
                password:function(){
                    console.log("正在输入密码...")
                }
            }
        })
    </script>
```

在 IE 11 浏览器里面运行，并打开控制台，然后在输入框中输入姓名和密码。可以发现，当每次输入时，都会调用对应的方法打印内容，如图 6-16 所示。

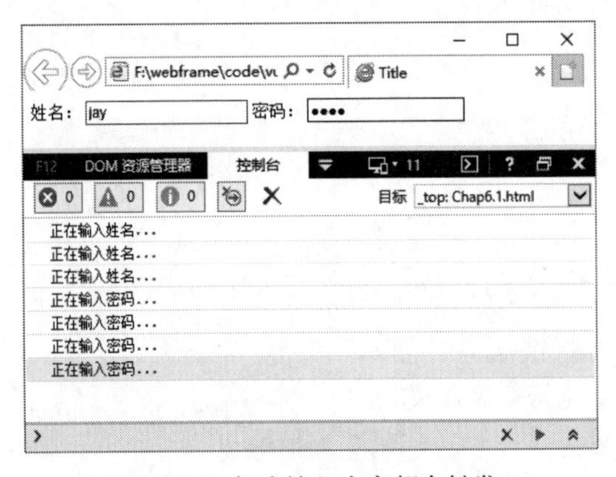

图 6-16　每次输入内容都会触发

在 Vue 中，提供了一种便利的方式去实现监听按键事件。在监听键盘事件时，经常需要查找常见的按键所对应的 keyCode，而 Vue 为最常用的按键提供了绝大多数常用的按键码的别名：

- .enter
- .tab
- .delete（捕获“删除”和“退格”键）
- .esc
- .space
- .up
- .down
- .left
- .right

对于上面的示例，每次输入都会触发 keyup 事件，有时候不需要每次输入都会触发，例如发 QQ 消息，希望所有的内容都输入完成再发送。这时可以为 keyup 事件添加 Enter 按键码，当键盘上的 Enter 键抬起时才会触发 keyup 事件。

例如，更改上面的示例，在 keyup 事件后添加 enter 按键码。

【例 6.13】添加 enter 按键码。

```
<div id="app">
    <label for="name">姓名: </label>
    <input v-on:keyup.enter="name" type="text" id="name">
</div>
<script>
    new Vue({
        el: '#app',
        methods: {
            name:function(){
                console.log("正在输入姓名...")
            }
        }
    })
</script>
```

在 IE 11 浏览器中运行，在 input 输入框中输入姓名（“金城武”），然后按下 Enter 键，弹起后触发 keyup 方法，打印“正在输入姓名…”，效果如图 6-17 所示。

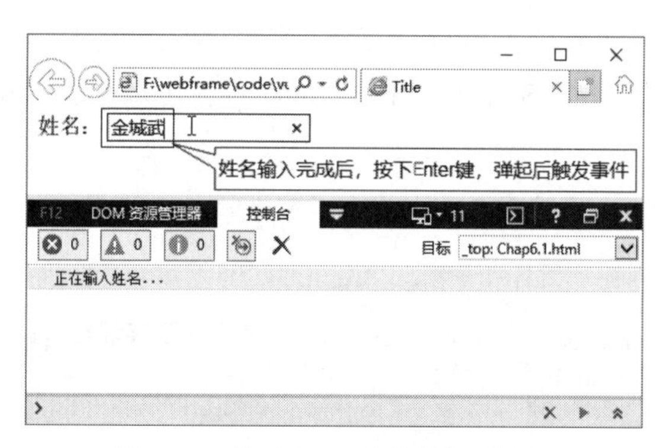

图 6-17　按下 Enter 键并弹起时触发

6.5 系统修饰键

可以用下面的修饰符来实现仅在按下相应按键时才触发鼠标或键盘事件的监听器。

- .ctrl
- .alt
- .shift
- .meta

注意

系统修饰键与常规按键不同，在和 keyup 事件一起用时，事件触发时修饰键必须处于按下状态。换句话说，只有在按住 ctrl 键的情况下释放其他按键，才能触发 keyup.ctrl。而单单释放 ctrl 键也不会触发事件。

【例 6.14】系统修饰键。

```
<div id="app">
    <label for="name">姓名: </label>
    <!--添加shift按键码-->
    <input v-on:keyup.shift.enter="name" type="text" id="name">
</div>
<script>
    new Vue({
        el: '#app',
        methods: {
            name:function(){
                console.log("正在输入姓名...")
            }
        }
    })
</script>
```

在 IE 11 浏览器中运行，在 input 中输入完成后，这时按下 Enter 键是无法激活 keyup 事件的，首先需要按下 Shift 键，再按 Enter 键才可以触发，效果如图 6-18 所示。

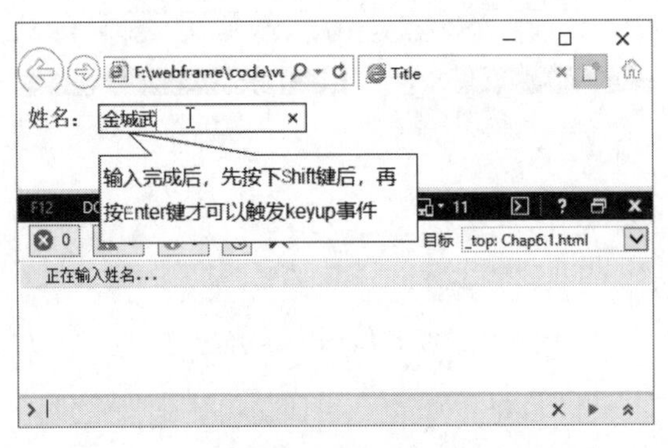

图 6-18　系统修饰键

6.6 案例实战——仿淘宝 Tab 栏切换

案例使用事件来定义 Tab 栏。

首先，为导航栏的每个 a 标签设置 curId 值，如果 curId=0，第一个 a 标签添加 cur 类名，如果 curId=1，第二个 a 标签添加 cur 类名，以此类推。添加了 cur 类名，a 标签就会改变样式。

然后根据 curId 的值显示 Tab 栏的内容，如果 curId=0，第一个 div 显示，其他三个 div 不显示。如果 curId=1，第二个 div 显示，其他三个 div 不显示，以此类推。

具体的实现代码如下：

```
<!DOCTYPE html>
<html lang="en">
<head>
    <meta charset="UTF-8">
    <title>Title</title>
    <script src="vue.js"></script>
</head>
<body>
<div id="tab">
    <div class="tab-tit">
        <a href="javascript:;" @click="curId=0" :class="{'cur':curId===0}">女装</a>
        <a href="javascript:;" @click="curId=1" :class="{'cur':curId===1}">鞋子</a>
        <a href="javascript:;" @click="curId=2" :class="{'cur':curId===2}">包包</a>
        <a href="javascript:;" @click="curId=3" :class="{'cur':curId===3}">男装</a>
    </div>
    <div class="tab-con">
        <div v-show="curId===0">
            <img src="001.jpg" alt="" width="100%">
        </div>
        <div v-show="curId===1">
            <img src="002.jpg" alt="" width="100%">
        </div>
        <div v-show="curId===2">
            <img src="003.jpg" alt="" width="100%">
        </div>
        <div v-show="curId===3">
            <img src="004.jpg" alt="" width="100%">
        </div>
    </div>
</div>
</body>
<script>
    new Vue({
        el: '#tab',
        data: {
            curId: 0
        }
    })
</script>
</html>
```

样式代码如下：

```
<style>
        #tab{
            width: 600px;
            margin: 0 auto;
        }
        .tab-tit{
            font-size: 0;
            width: 600px;
        }
        .tab-tit a{
```

```
            display: inline-block;
            height: 40px;
            line-height: 40px;
            font-size: 16px;
            width: 25%;
            text-align: center;
            background: #e1e1e1;
            color: #333;
            text-decoration: none;
        }
        .tab-tit .cur{
            background: #09f;
            color: #fff;
        }
        .tab-con div{
            border: 1px solid #e7e7e7;
            height: 400px;
            padding-top: 20px;
        }
    </style>
```

在 IE 浏览器中运行，效果如图 6-19 所示。

图 6-19　Tab 栏切换效果

6.7 疑难解惑

疑问 1：exact 修饰符如何使用？

exact 修饰符允许控制由精确的系统修饰符组合触发的事件。

```
<!--即使Alt键或Shift键被一同按下时也会触发-->
<button @click.ctrl="onClick">A</button>
<!--有且只有Ctrl键被按下的时候才触发-->
<button @click.ctrl.exact="onCtrlClick">A</button>
<!--没有任何系统修饰符被按下的时候才触发-->
<button @click.exact="onClick">A</button>
```

疑问 2：为什么在 HTML 中监听事件？

可能注意到这种事件监听的方式违背了关注点分离（separation of concern）传统的理念。不必担心，因为所有的 Vue 事件处理方法和表达式都严格绑定在当前视图的 ViewModel 上，它不会导致任何维护上的困难。实际上，使用 v-on 有以下几个好处。

（1）扫一眼 HTML 模板便能轻松定位在 JavaScript 代码里对应的方法。

（2）因为无须在 JavaScript 里手动绑定事件， ViewModel 代码可以是非常纯粹的逻辑，和 DOM 完全解耦，更易于测试。

（3）当一个 ViewModel 被销毁时，所有的事件处理器都会自动被删除。无须担心如何清理它们。

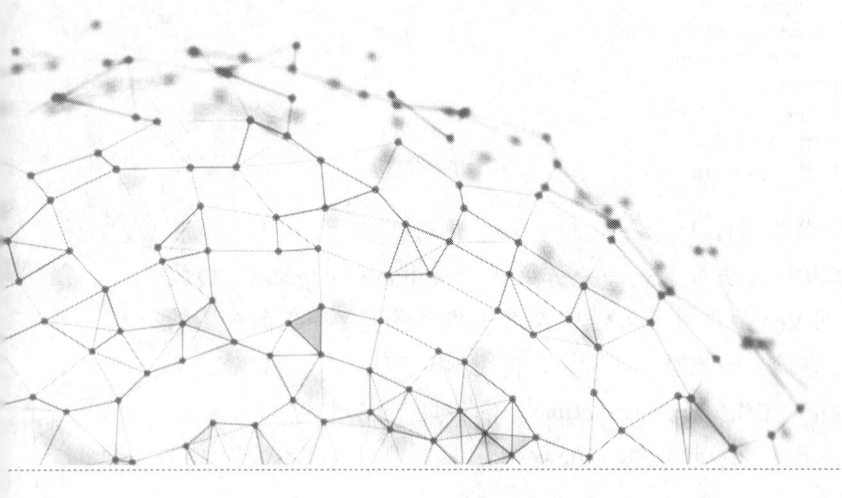

第7章

表单输入绑定（双向数据绑定）

对于 Vue 来说，使用 v-bind 并不能解决表单域对象双向绑定的需求，所谓双向绑定，就是无论是通过 input 还是通过 Vue 对象，都能修改绑定的数据对象的值。Vue 提供了 v-model 进行双向绑定。

7.1 双向绑定

数据的绑定，不管是使用插值表达式（{{}}）还是 v-text 指令，对于数据间的交互都是单向的，只能将 Vue 实例里的值传递给页面，页面对数据值的任何操作却无法传递给 model。

MVVM 模式最重要的一个特性是数据的双向绑定，而 Vue 作为一个 MVVM 框架，也实现了数据的双向绑定。在 Vue 中使用内置的 v-model 指令完成数据在 View 与 Model 间的双向绑定。

可以用 v-model 指令在表单 <input>、<textarea> 及 <select> 元素上创建双向数据绑定。它会根据控件类型自动选取正确的方法来更新元素。尽管有些神奇，但 v-model 本质上不过是语法糖。它负责监听用户的输入事件以更新数据，并对一些极端场景进行特殊处理。

v-model 会忽略所有表单元素的 value、checked、selected 特性的初始值，而总是将 Vue 实例的数据作为数据来源。我们应该通过 JavaScript 在组件的 data 选项中声明初始值。

提示

表单元素可以与用户进行交互，所以使用 v-model 指令在表单控件或者组件上创建双向绑定。

7.2 基本用法

v-model其实是相当于把Vue中的属性绑定到元素（input）上，如果该数据属性有值，值会显示到input中，同时元素中输入的内容也决定Vue中的属性值。

v-model在内部为不同的输入元素，使用不同的属性并抛出不同的事件：

- text和textarea元素使用value属性和input事件。
- checkbox和radio使用checked属性和change事件。
- select字段将value作为prop并将change作为事件。

注意

对于需要使用输入法（如中文、日文、韩文等）的语言，会发现v-model不会在输入法组合文字过程中得到更新。如果也想处理这种情况，可使用input事件。

7.2.1 文本

在下面的示例中，绑定了name和age两个属性。

【例7.1】绑定文本。

```
<div id="app">
    <label for="name">姓名: </label>
    <input v-model="name" type="text" id="name">
    <p>{{name}}</p>
    <label for="age">年龄: </label>
    <input v-model="age" type="text" id="age">
    <p>{{age}}</p>
</div>
<script>
    new Vue({
        el: '#app',
        data:{
            name:'金城武',
            age:'40'
        }
    })
</script>
```

在IE 11浏览器里面运行，效果如图7-1所示；把姓名改为“老王”，年龄改为“50”，p标签中的内容也随着改变，如图7-2所示。

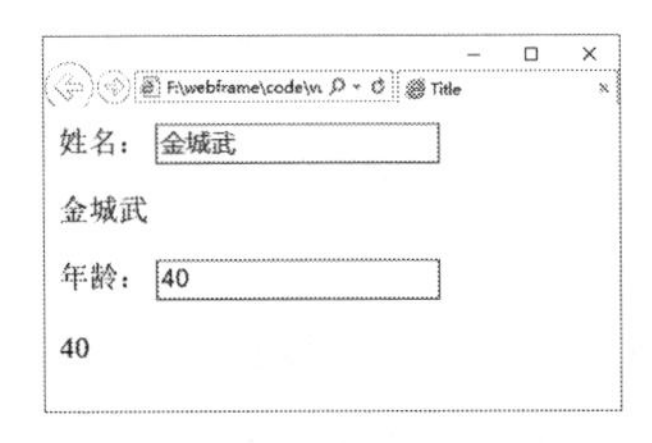

图7-1　页面初始化效果

图7-2　变更后的效果

7.2.2 多行文本

把上面示例中的 p 标签换成 textarea 标签，即可实现多行文本的绑定。

【例 7.2】绑定多行文本。

```
<div id="app">
    <span>第一章内容:</span>
    <p style="white-space: pre-line;">{{ message }}</p>
    <textarea v-model="message"></textarea>
</div>
<script>
    new Vue({
        el: '#app',
        data:{
            message:""
        }
    })
</script>
```

在 IE 11 浏览器里面运行，在 textarea 标签中输入多行文本，效果如图 7-3 所示。

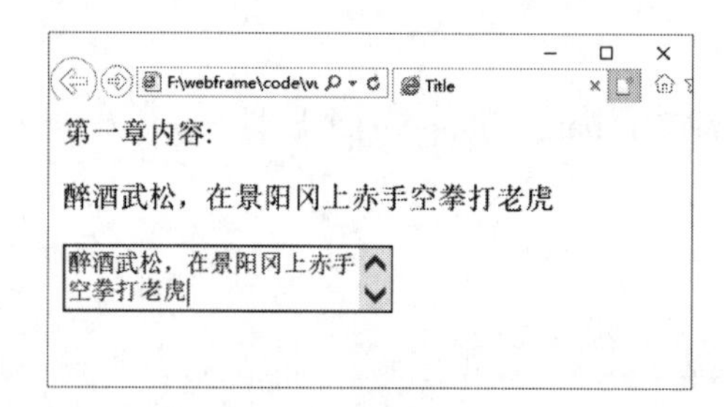

图 7-3　绑定多行文本

7.2.3 复选框

单个复选框，绑定到布尔值。

【例 7.3】绑定单个复选框。

```
<div id="app">
    <input type="checkbox" id="checkbox" v-model="checked">
    <label for="checkbox">{{ checked }}</label>
</div>
<script>
    new Vue({
        el: '#app',
        data:{
            checked:false
        }
    })
</script>
```

在 IE 11 浏览器中运行，效果如图 7-4 所示；当选中复选框后，效果如图 7-5 所示。

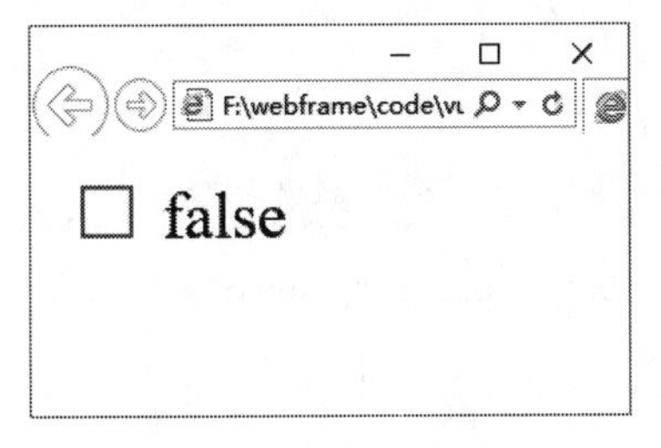

图 7-4 页面初始化效果

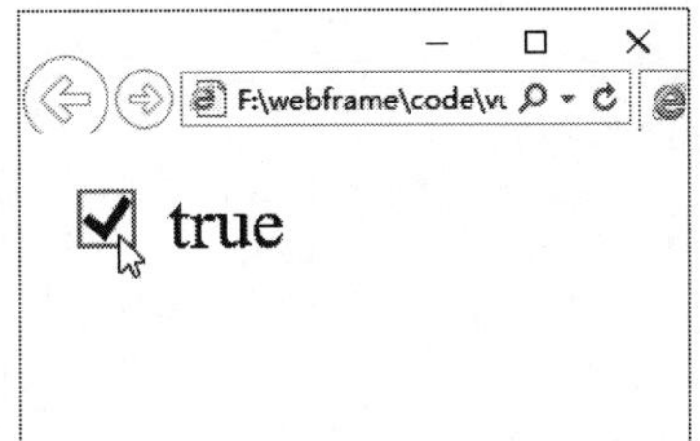

图 7-5 选中效果

多个复选框，绑定到同一个数组，被选中的添加到数组中。

【例 7.4】绑定多个复选框。

```
<div id="app">
    <input type="checkbox" id="name1" value="马云" v-model="checkedNames">
    <label for="name1">马云</label>
    <input type="checkbox" id="name2" value="马化腾" v-model="checkedNames">
    <label for="name2">马化腾</label>
    <input type="checkbox" id="name3" value="马明哲" v-model="checkedNames">
    <label for="name3">马明哲</label>
    <p><span>哪些人做过首富: {{ checkedNames }}</span></p>
</div>
<script>
    new Vue({
        el: '#app',
        data:{
            checkedNames: [],    //定义空数组
        }
    })
</script>
```

在 IE 11 浏览器里面运行，选中前两个复选框，选中的内容显示在数组中，如图 7-6 所示。

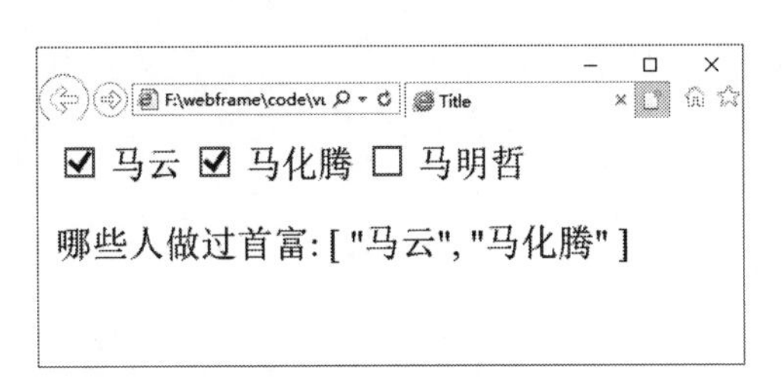

图 7-6 绑定多个复选框

7.2.4 单选按钮

单选按钮一般都有多个条件可供选择，既然是单选按钮，自然希望实现互斥效果，可以使用 v-model 指令配合单选按钮的 value 来实现。

在下面的示例中，多个单选按钮绑定到同一个数组，被选中的添加到数组中。

【例 7.5】绑定单选按钮。

```
<div id="app">
    <h3>单选题</h3>
    <input type="radio" id="one" value="A" v-model="picked">
    <label for="one">A</label><br/>
    <input type="radio" id="two" value="B" v-model="picked">
    <label for="two">B</label><br/>
    <input type="radio" id="three" value="C" v-model="picked">
    <label for="three">C</label><br/>
    <input type="radio" id="four" value="D" v-model="picked">
    <label for="four">D</label>
    <p><span>选择: {{ picked }}</span></p>
</div>
<script>
    new Vue({
        el: '#app',
        data:{
            picked: ''
        }
    })
</script>
```

在 IE 11 浏览器里面运行，选中 B 选项，效果如图 7-7 所示。

图 7-7 绑定单选按钮

7.2.5 选择框

1. 单选框

绑定单选框的案例如下。

【例 7.6】绑定单选框。

```
<div id="app">
    <h3>选择你最喜欢吃的水果</h3>
    <select v-model="selected">
        <option disabled value="">可以选择的水果如下</option>
        <option>苹果</option>
        <option>香蕉</option>
        <option>橘子</option>
    </select>
    <span>选择结果: {{ selected }}</span>
</div>
<script>
    new Vue({
```

```
        el: '#app',
        data:{
            selected: ' '
        }
    })
</script>
```

在 IE 11 浏览器里面运行，结果如图 7-8 所示。

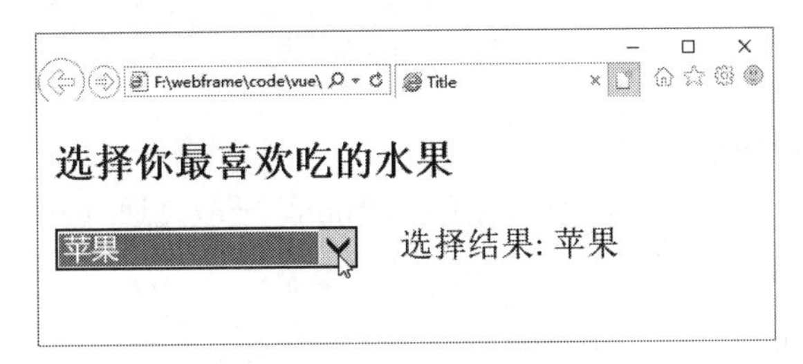

图 7-8　绑定单选框

提示

如果 v-model 表达式的初始值未能匹配任何选项，<select> 元素将被渲染为“未选中”状态。在 iOS 中，这会使用户无法选择第一个选项。因为这样的情况下，iOS 不会触发 change 事件。因此，推荐像上面这样提供一个值为空的禁用选项。

2. 多选框（绑定到一个数组）

为 <select> 标签添加 multiple 属性，即可实现多选。

【例 7.7】绑定多选框。

```
<div id="app">
    <h3>选择你喜欢吃的水果</h3>
    <select v-model="selected" multiple style="height: 100px">
        <option disabled value="">可以选择的水果如下</option>
        <option>苹果</option>
        <option>香蕉</option>
        <option>橘子</option>
        <option>草莓</option>
    </select>
    <span>选择结果: {{ selected }}</span>
</div>
<script>
    new Vue({
        el: '#app',
        data:{
            selected: [],
        }
    })
</script>
```

在 IE 11 浏览器里面运行，选择多个选项，效果如图 7-9 所示。

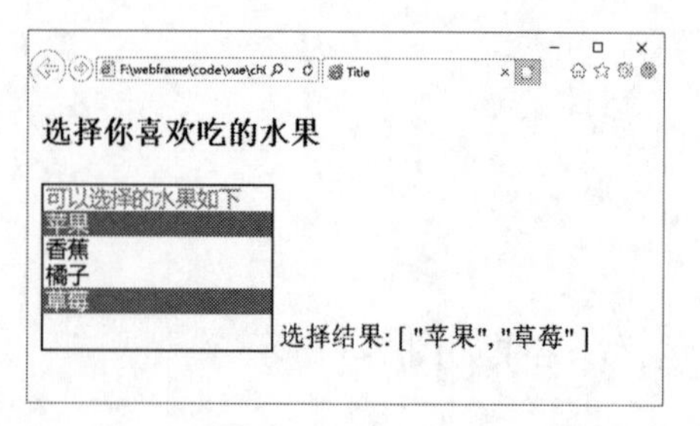

图 7-9　绑定多选框

3. 用 v-for 渲染的动态选项

在实际应用场景中，<select> 标签中的 <option> 一般是通过 v-for 指令动态输出的，其中每一项的 value 或 text 都可以使用 v-bind 动态输出。

【例 7.8】用 v-for 渲染的动态选项。

```
<div id="app">
    <select v-model="selected">
        <option v-for="option in options" v-bind:value="option.value">
        <option>{{option.text}}</option>
    </select>
    <span>选择结果: {{ selected }}</span>

</div>
<script>
    new Vue({
        el: '#app',
        data: {
            selected: '苹果',
            options: [
                { text: 'One', value: '苹果' },
                { text: 'Two', value: '香蕉' },
                { text: 'Three', value: '杧果' }
            ]
        }
    })
</script>
```

在 IE 11 浏览器里面运行，在选择框中选择 Three，效果如图 7-10 所示。

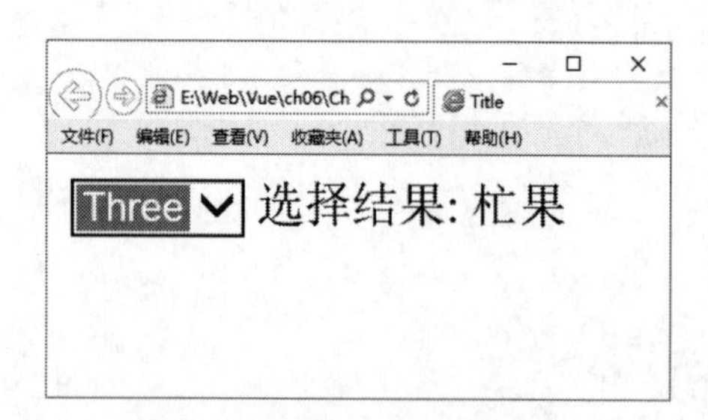

图 7-10　v-for 渲染的动态选项

7.3 值绑定

对于单选按钮，复选框及选择框的选项，v-model 绑定的值通常是静态字符串 (对于复选框也可以是布尔值)。但是有时可能想把值绑定到 Vue 实例的一个动态属性上，这时可以用 v-bind 实现，并且这个属性的值可以不是字符串。

7.3.1 绑定复选框

在下面示例中，true-value 和 false-value 特性并不会影响输入控件的 value 特性，因为浏览器在提交表单时并不会包含未被选中的复选框。如果要确保表单中这两个值中的一个能够被提交，(例如 yes 或 no)，请换用单选按钮。

【例 7.9】动态绑定复选框。

```
<div id="app">
    <input type="checkbox" v-model="toggle" true-value="yes" false-value="no">
    <span>{{toggle}}</span>
</div>
<script>
    new Vue({
        el: '#app',
        data: {
            toggle:'',
        }
    })
</script>
```

在 IE 11 浏览器里面运行，选中状态如图 7-11 所示，反之如图 7-12 所示。

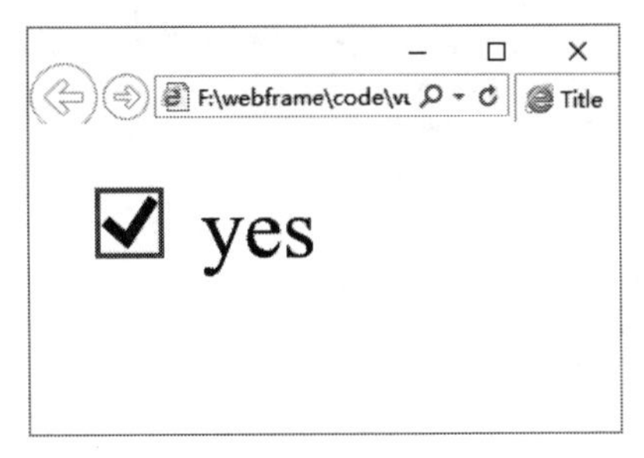

图 7-11　选中效果

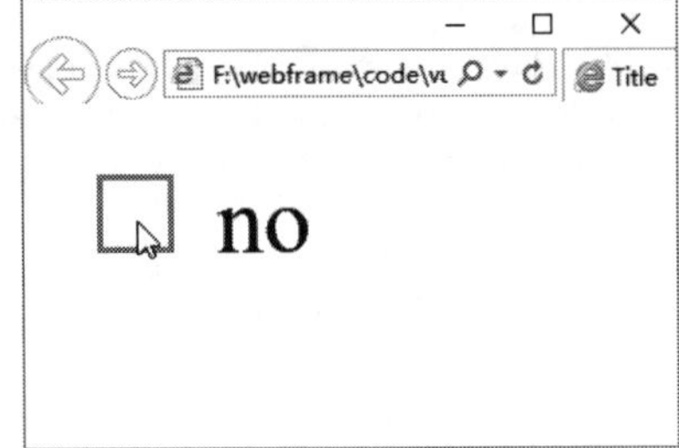

图 7-12　不选中效果

7.3.2 绑定单选按钮

首先为单选按钮绑定一个属性 a，定义属性值为“苹果”；然后使用 v-model 指令为单选按钮绑定 pick 属性，当单选按钮选中后，pick 的值等于 a 的属性值。

【例 7.10】动态绑定单选按钮的值。

```
<div id="app">
    <input type="radio"  v-model="pick" v-bind:value="a">
    <span>{{ pick}}</span>
</div>
<script>
    new Vue({
        el: '#app',
        data: {
            a:'该单选按钮已被选中',
            pick:'',
        }
    })
</script>
```

在 IE 11 浏览器里面运行，选中效果如图 7-13 所示。

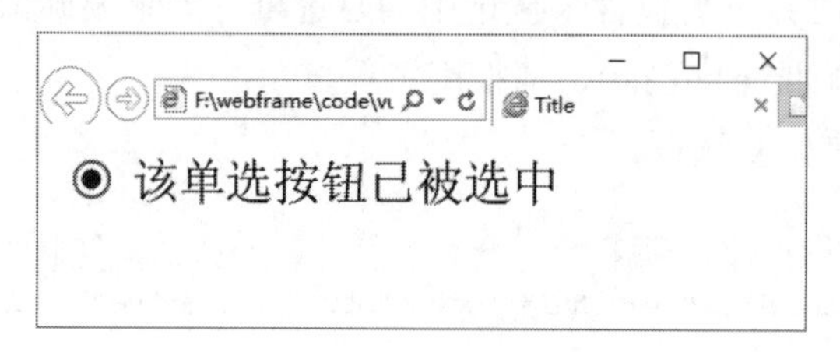

图 7-13　单选按钮选中效果

7.3.3　绑定选择框

在下面的示例中，定义了 4 个 option 选项，使用 v-bind 进行绑定。

【例 7.11】动态绑定选择框的选项。

```
<div id="app">
    <select v-model="selected" multiple>
        <option v-bind:value="{ number: 1 }">A</option>
        <option v-bind:value="{ number: 2 }">B</option>
        <option v-bind:value="{ number: 3 }">C</option>
        <option v-bind:value="{ number: 4 }">D</option>
    </select>
    <p><span>{{ selected }}</span></p>
</div>
<script>
    new Vue({
        el: '#app',
        data: {
            selected:[],

        }
    })
</script>
```

在 IE 11 浏览器里面运行，选中 C 选项，在 p 标签中显示相应的 number 值，如图 7-14 所示。

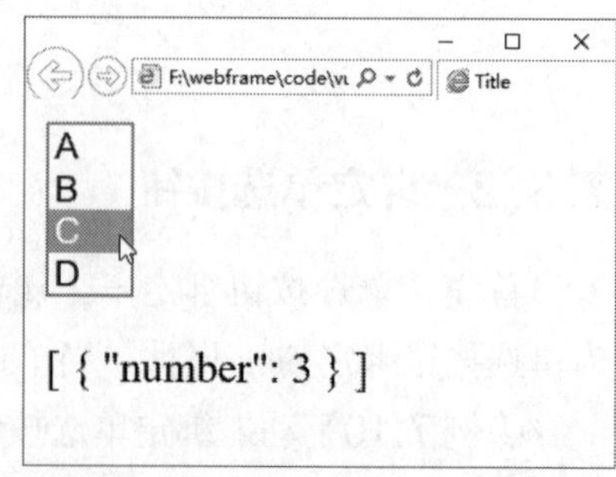

图 7-14　动态绑定选择框的选项

7.4 修饰符

对于 v-model 指令，还有 3 个常用的修饰符：lazy、number 和 trim。下面分别介绍。

7.4.1 lazy 修饰符

在输入框中，v-model 默认是同步数据，使用 lazy 修饰符会转变为在 change 事件中同步，也就是在失去焦点或者按下 Enter 键时才更新。

【例 7.12】lazy 修饰符示例。

```
<div id="app">
    <input v-model.lazy="message">
    <span>{{ message }}</span>
</div>
<script>
    new Vue({
        el: '#app',
        data: {
            message:'abc',
        }
    })
</script>
```

在 IE 11 浏览器里面运行，输入“abc123456789”，如图 7-15 所示；失去焦点后同步数据，如图 7-16 所示。

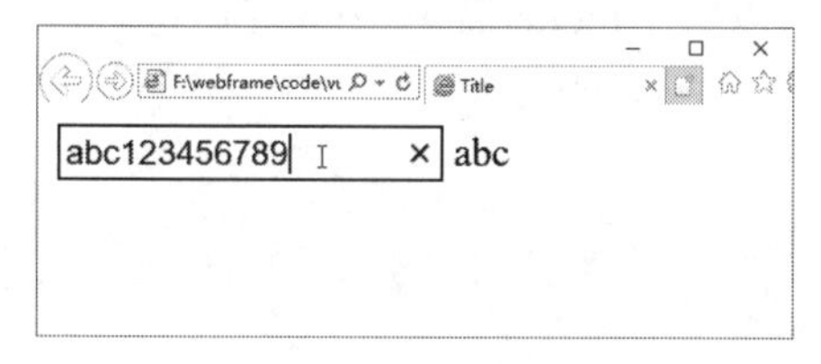

图 7-15　输入数据

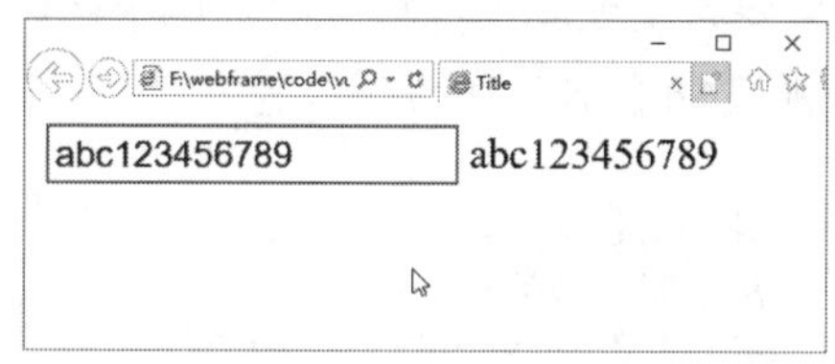

图 7-16　失去焦点同步数据

7.4.2 number 修饰符

number 修饰符可以将输入的值转化为 Number 类型，否则虽然输入的是数字，但它的类型其实是 String，此修饰符在数字输入框中比较有用。

如果想自动将用户的输入值转为数值类型，可以给 v-model 添加 number 修饰符，因为即使在 type="number" 时，HTML 输入元素的值也总会返回字符串。如果这个值无法被 parseFloat() 解析，则会返回原始的值。

【例 7.13】number 修饰符。

```
<div id="app">
        <p>.number修饰符</p>
```

```
        <input type="number" v-model.number="val">
        <p>数据类型是: {{ typeof(val) }}</p>
</div>
<script>
    new Vue({
        el: '#app',
        data:{
            val:'',
        }
    })
</script>
```

在 IE 11 浏览器里面运行，输入“123456789”，由于使用了 number 修饰符，所以显示的数据类型为 number 类型，如图 7-17 所示。

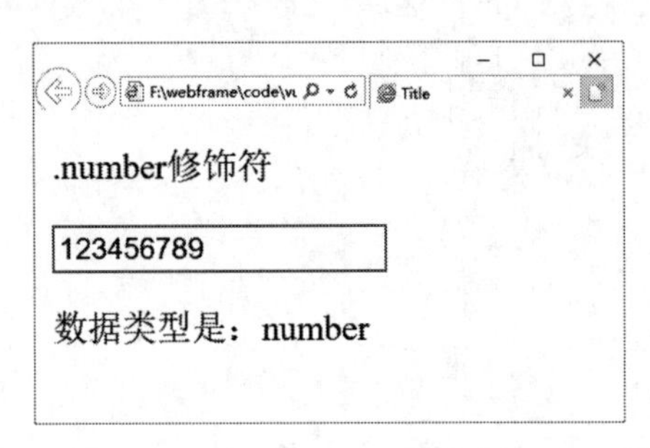

图 7-17　number 修饰符

7.4.3　trim 修饰符

如果要自动过滤用户输入的首尾空格，可以给 v-model 添加 trim 修饰符。

【例 7.14】 trim 修饰符示例。

```
<div id="app">
    <p>.trim修饰符</p>
    <input type="text" v-model.trim="val">
    <p>val的长度是: {{ val.length }}</p>
</div>
<script>
    new Vue({
        el: '#app',
        data:{
            val:'',
        }
    })
</script>
```

在 IE 11 浏览器里面运行，在 input 前后输入了许多空格，并输入“123”，可以看到 val 的长度为 3，效果如图 7-18 所示。

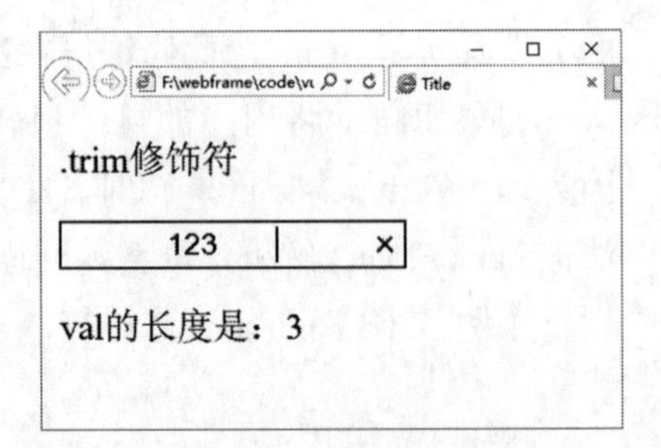

图 7-18　trim 修饰符示例

7.5 案例实战1——小游戏破坏瓶子

本案例应用前面所学的知识，来编写一个简单的破坏瓶子的小游戏。小游戏界面中放入一张图片，设置一些按钮，不断地单击按钮，当单击一定次数后，用一张新图片替换原来的图片。

在IE11浏览器中运行，效果如图7-19所示；当我们不断单击“敲瓶子”按钮后，瓶子会被破坏，效果如图7-20所示。

图7-19　完整的瓶子

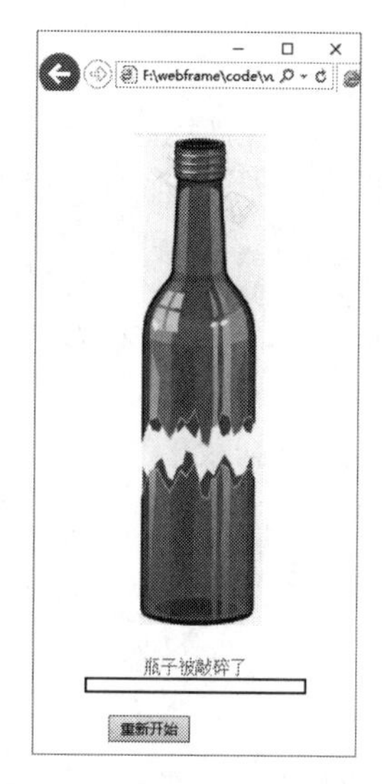

图7-20　被敲碎后的瓶子

下面来看一下实现的步骤。

第一步：先来看一下HTML结构。结构包括图片、提示信息、破坏进度和控制按钮。

```
<div id="app">
    <!--图片-->
    <div id="bottle" v-bind:class="{ burst:boole }"></div>
    <!--提示破碎的信息-->
    <div id="state">{{ state }}</div>
    <!--破坏进度情况-->
    <div id="bottle-health">
        <div v-bind:style="{width:health + '%' }"></div>
    </div>
    <!--控制按钮-->
    <div id="controls">
        <button v-on:click="beat" v-show="!boole">敲瓶子</button>
        <button v-on:click="restart">重新开始</button>
    </div>
</div>
```

第二步：设计样式。

```
<style>
```

```
        #bottle{
            width:150px;                  /*定义宽度*/
            height: 500px;                /*定义高度*/
            margin: 0 auto;               /*定义外边距*/
            background: url(001.png) center no-repeat;   /* 定义图片 居中 不平铺*/
            background-size: 80%;          /*定义背景图片尺寸*/
        }
        #bottle.burst{
            background-image:url(002.png);      /* 定义背景图片*/
        }
        #state{
            color: red;                         /*定义字体颜色*/
            text-align: center;                 /*定义水平居中*/
        }
        #bottle-health{
            width:200px;                        /*定义宽度*/
            border: 2px solid #000;             /*定义边框*/
            margin: 0 auto 20px auto;           /*定义外边距*/
        }
        #bottle-health div{
            height:10px;                        /*定义高度*/
            background: #dc2b57;                /*定义背景颜色*/
        }
        #controls{
            width: 200px;                       /*定义宽度*/
            margin: 0 auto;                     /*定义外边距*/
        }
        #controls button{
            margin-left: 20px;                  /*定义左侧外边距*/
        }
</style>
```

第三步，设计 Vue 逻辑。

```
<script>
    new Vue({
        el: '#app',
        data:{
            health:100,    //定义破坏的进度条
            boole:false,
            state:""
        },
        methods:{
            //破坏瓶子
            beat:function () {
                this.health-=10;      //每次单击按钮触发beat方法，health宽度减小10%
                if (this.health<=0){
                this.boole=true;      //当this.health<=0时，this. boole的值为
                                      //true，应用burst类更换背景图片
                    this.state="瓶子被敲碎了"   //提示“瓶子被敲碎”信息
                }
            },
            //重新开始
            restart:function(){
                this.health=100;      //进度条恢复100
                this.boole=false;     //this.boole回到定义时的值（false），显示瓶子
                                      //没破坏前图片
            }
```

```
        }
    })
</script>
```

7.6 案例实战 2——设计动态表格

本案例设计了一个动态表格，可以手动增加、删除和更新数据。数据的添加和更新使用双向数据绑定来实现。

首先设计页面结构，代码如下：

```
<!DOCTYPE html>
<html>
<head>
    <meta charset="utf-8">
    <script src="vue.min.js"></script>
    <title>动态表格</title>
<body>
<div id="table">
    <div class="add">
        <input type="text" v-model="addDetail.title" name="title" placeholder=
            "发布内容" />
        <input type="text" v-model="addDetail.user" name="user" placeholder=
            "发布人" />
        <input type="date" v-model="addDetail.dates" name="date" placeholder=
            "发布时间" />
        <button @click="add">新增</button>
    </div>
    <table cellpadding="0" cellspacing="0">
        <thead>
        <tr>
            <th>序号</th>
            <th>标题</th>
            <th>发布人</th>
            <th>发布时间</th>
            <th>操作</th>
        </tr>
        </thead>
        <tbody>
        <tr v-for="(item,index) in newsList">
            <td width="10%">{{index+1}}</td>
            <td>{{item.title}}</td>
            <td width="15%">{{item.user}}</td>
            <td width="20%">{{item.dates}}</td>
            <td width="15%">
                <span @click="deletelist(item.id,index)" class="delete">删除</span>
                <span class="edit" @click="edit(item)">编辑</span>
            </td>
        </tr>
        </tbody>
    </table>
    <div id="mask" v-if="editlist">
```

```
            <div class="mask">
                <div class="title">
                    编辑
                    <span @click="editlist=false">x</span>
                </div>
                <div class="content">
                        <input type="text" v-model="editDetail.title" name="title"
                          value="" placeholder="标题" />
                         <input type="text" v-model="editDetail.user" name="user"
                          value="" placeholder="发布人" />
                        <input type="date" v-model="editDetail.dates" name="date"
                          value="" placeholder="发布时间" />
                    <button @click="update">更新</button>
                    <button @click="editlist=false">取消</button>
                </div>
            </div>
        </div>
    </div>
    </body>
    </html>
```

样式代码如下：

```
#table table {width: 100%;font-size: 14px;border: 1px solid #eee;}
        #table {padding: 0 10px;}
        table thead th {background: #f5f5f5;padding: 10px;text-align: left;}
        table tbody td {
             padding: 10px;text-align: left;border-bottom: 1px solid
               #eee;border-right: 1px solid #eee;
        }
        table tbody td span{margin: 0 10px;cursor: pointer;}
        .delete {color: red;}
        .edit {color: #008cd5;}
        .add {border: 1px solid #eee;margin: 10px 0;padding: 15px;}
        input {
               border: 1px solid #ccc;padding: 5px;border-radius: 3px;margin-
                 right: 15px;
        }
        button {
               background: #008cd5;border: 0;padding: 4px 15px;border-radius:
                 3px;color: #fff;
        }
        #mask {
              background: rgba(0, 0, 0, .5);width: 100%;height: 100%;position:
                 fixed;z-index: 4;top: 0;left: 0;
        }
        .mask {
                 width: 300px;height: 250px;background: rgba(255, 255, 255,
                   1);position: absolute;left: 0;
              top: 0;right: 0;bottom: 0;margin: auto;z-index: 47;border-radius:
                5px;
        }
        .title {
            padding: 10px;border-bottom: 1px solid #eee;
        }
        .title span {float: right;cursor: pointer;}
        .content {padding: 10px;}
        .content input {width: 270px;margin-bottom: 15px;}
```

在 JavaScript 中，定义数据，并设置两条数据；定义新增、删除、编辑和更新等方法，通过单击事件触发。

```
<script>
    var app = new Vue({
        el: '#table',
        data: {
            addDetail:{},
            editlist: false,
            editDetail: {},
            newsList: [{
                title: '招聘前端工程师',
                user: '关羽',
                dates: '2020-08-10',
                id: "10001"
            }, {
                title: '招聘PHP工程师',
                user: '张飞',
                dates: '2020-08-15',
                id: "10002"
            }],
            editid:''
        },
        mounted() {},
        methods: {
            //新增功能
            add:function() {
                this.newsList.push({
                    title: this.addDetail.title,
                    user: this.addDetail.user,
                    dates: this.addDetail.dates,
                })
            },
            //删除功能
            deletelist:function(id, i){
                this.newsList.splice(i, 1);
            },
            //编辑功能
            edit:function(item){
                this.editDetail = {
                    title: item.title,
                    user: item.user,
                    dates: item.dates,
                    id: item.id
                }
                this.editlist = true
                this.editid = item.id
            },
            //确认更新
            update:function() {
                let _this= this
                for(let i = 0; i < _this.newsList.length; i++) {
                    if(_this.newsList[i].id ==this.editid) {
                        _this.newsList[i] = {
                            title: _this.editDetail.title,
                            user: _this.editDetail.user,
                            dates: _this.editDetail.dates,
                            id: this.editid
```

```
                    }
                    this.editlist = false;
                }
            }
        }
    }
})
</script>
```

在谷歌浏览器中运行，在输入框中输入发布的内容、发布人，选择相应的日期，然后单击“新增”按钮，数据将添加到列表中，效果如图 7-21 所示。

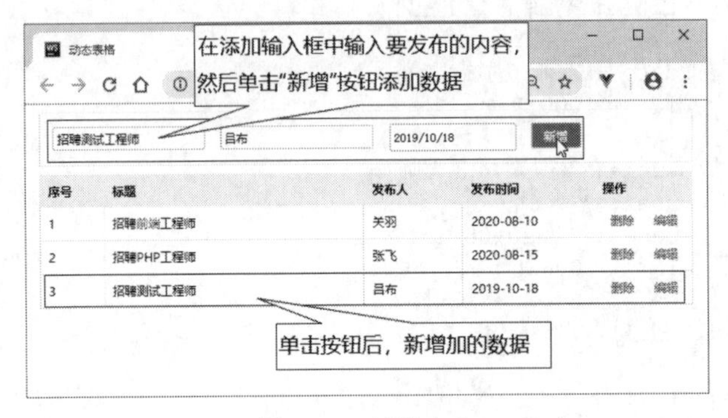

图 7-21　新增内容

单击“删除”按钮，可删除对应的数据；单击“编辑”按钮，将进入更新界面，在更新界面中可以更改数据，然后单击“更新”按钮更新数据。页面效果如图 7-22 所示。

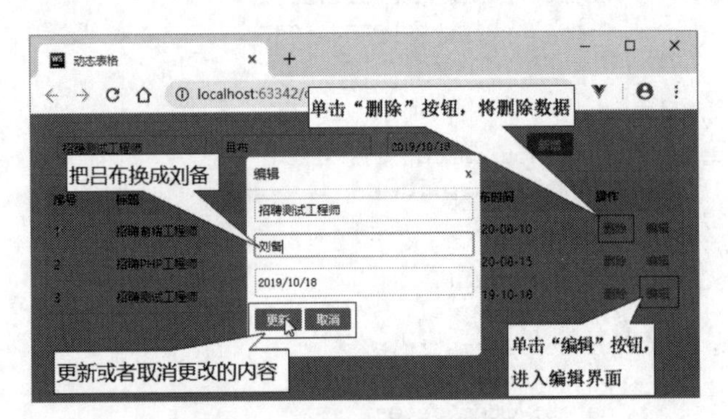

图 7-22　删除、编辑和更新效果

7.7 疑难解惑

疑问 1：v-model 是什么？Vue 中标签怎么绑定事件？

Vue 中利用 v-model 进行表单数据的双向绑定。具体做了以下两个操作。

（1）v-bind 绑定了一个 value 的属性。

（2）利用 v-on 把当前的元素绑定到一个事件上。

例如下面的示例：

```
<div id="app">
    <!--绑定value属性，input绑定到oninput事件上-->
    <input v-model:value="inputValue" v-on:input="inputValue=$event.target.value">
    <p>----{{inputValue}}----</p>
</div>
<script>
    new Vue({
        el:"#app",
        data:{
            inputValue:""
        }
    })
</script>
```

在谷歌浏览器中运行，效果如图 7-23 所示。

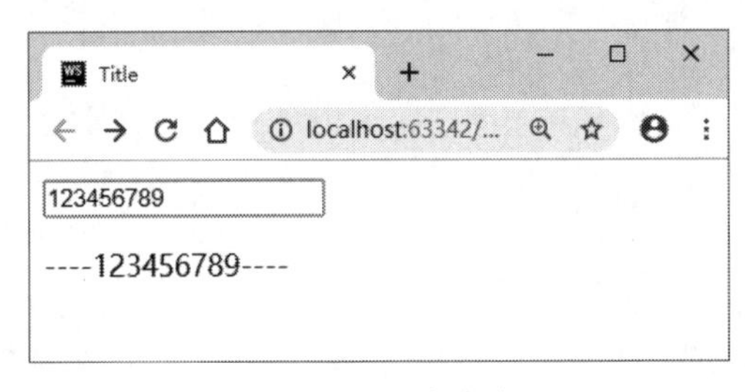

图 7-23　双向绑定

input 元素本身有个 oninput 事件，这是 HTML 5 新增加的，类似 onchange，每当输入框内容发生变化时，就会触发 oninput，把最新的 value 传递给 inputValue。

疑问 2：如何在组件上使用双向数据绑定？

首先在模板中定义数据双向绑定，然后通过 props 属性把 price 传递给 myon-input 组件。例如下面代码：

```
<div id="app">
    <myon-input v-model="price"></myon-input>
    <span>{{price}}</span>
</div>
<script>
    //定义组件myon-input
    Vue.component('myon-input', {
        template: '
                <span>
                  <input
                    :value="value"
                    @input="$emit('input', $event.target.value)">
                </span>
              ',
         props: ['value'],    //向组件传递price的值
    })
new Vue({
  el: '#app',
```

```
    data: {
      price:1,
    }
  })
</script>
```

在谷歌浏览器运行，效果如图 7-24 所示。

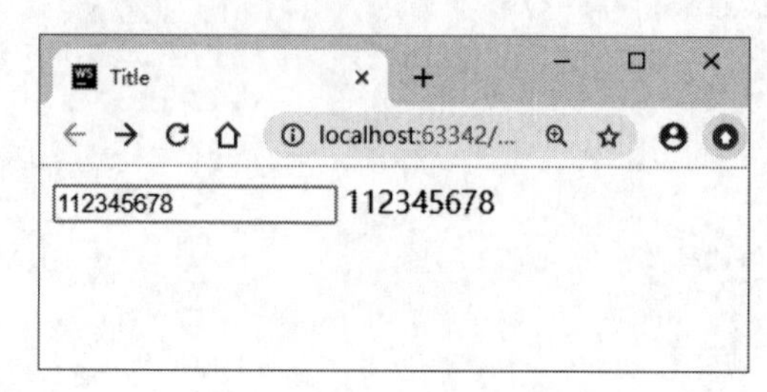

图 7-24　在组件上实现双向绑定

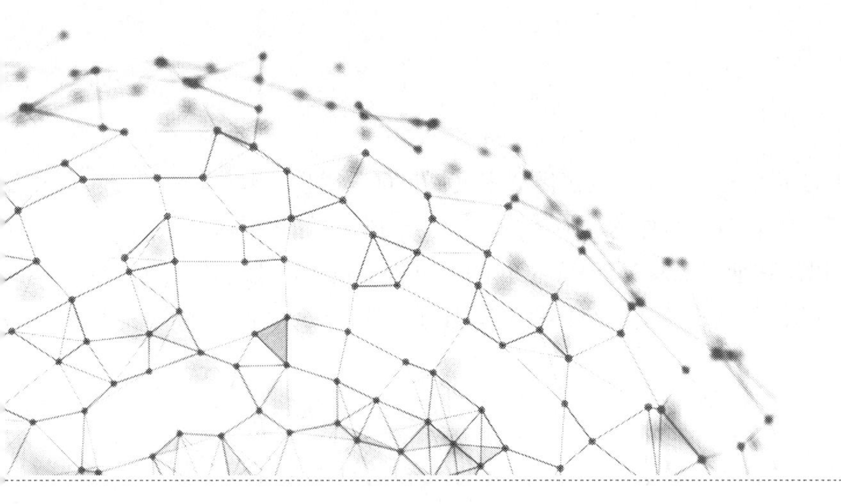

第8章 组件技术

在之前的章节中，对于 Vue 的一些基础语法进行了简单的讲解，通过之前的代码可以清晰地看出，使用 Vue 的整个过程，最终都是在对 Vue 实例进行的一系列操作。这里就会引出一个问题，把所有对于 Vue 实例的操作全部写在一块，这必然会导致这个方法又长又不好理解。那么，如何在 Vue 中解决上述问题呢？这里就需要用到组件这个技术了，本章就来学习 Vue 中组件的知识。

8.1 组件是什么

组件是 Vue 中的一个重要概念，它是一个可以复用的 Vue 实例，拥有独一无二的组件名称，可以扩展 HTML 元素，以组件名称的方式作为自定义的 HTML 标签。因为组件是可复用的 Vue 实例，所以它们与 new Vue() 接收相同的选项，例如 data、computed、watch、methods 以及生命周期钩子等。仅有的例外是像 el 这样根实例特有的选项。

例如，在绝大多数的系统网页中，网页都包含 header、menu、body、footer 等部分，在很多时候，同一个系统中的多个页面，可能仅仅是页面中 body 部分显示的内容不同，因此，我们就可以将系统中重复出现的页面元素设计成一个个的组件，当需要使用到的时候，引用这个组件即可。

模块化主要是为了实现每个模块、方法的功能单一，一般是通过代码逻辑的角度进行划分；而 Vue 中的组件化，更多的是为了实现对于前端 UI 组件的重用。

8.2 组件的注册

在 Vue 中创建一个新的组件之后，为了能在模板中使用，这些组件必须先进行注册，以便 Vue 能够识别。在 Vue 中有两种组件的注册类型：全局注册和局部注册。

全局注册的组件，可以用在通过 new Vue() 新创建的 Vue 根实例中，也可以在组件树中的所有子组件的模板中使用；而局部注册的组件只能在当前注册的 Vue 实例中进行使用。

8.2.1 全局注册

在 Vue 中创建全局组件，通常的做法是先使用 Vue.extend 方法构建模板对象，然后通过 Vue.component 方法来注册组件。因为组件最后会被解析成自定义的 HTML 代码，因此，可以直接在 HTML 中使用组件名称作为标签来使用。

【例 8.1】全局注册组件。

```
<div id="app">
    <my-component></my-component>
</div>
<script>
    //1、使用 Vue.extend 构建模板对象
    var comElement = Vue.extend({
        template: '<div><h3>全局组件</h3><p>这是我们创建的全局组件</p></div>'
    })
    //2、使用 Vue.component 注册全局组件
    Vue.component('my-component', comElement)
    var app = new Vue({
        el: '#app'
    });
</script>
```

在谷歌浏览器里面运行，效果如图 8-1 所示。

图 8-1　全局注册组件

从控制台中可以看到，自定义的组件已经被解析成了HTML元素。需要注意一个问题，当采用小驼峰（myCom）的方式命名组件时，在使用这个组件的时候，需要将大写字母改成小写字母，同时两个字母之间需要使用“-”进行连接，例如<my-com>。

【例8.2】组件命名示例。

```
<div id="app">
    <!--小驼峰命名的组件，使用方式-->
    <my-com></my-com>
</div>
<script>
    var comElement = Vue.extend({
        template: '<h3>组件名称使用的规则</h3>'
    })
    //小驼峰命名组件
    Vue.component('myCom', comElement)
    var app = new Vue({
        el: '#app'
    });
</script>
```

在谷歌浏览器里面运行，效果如图8-2所示。

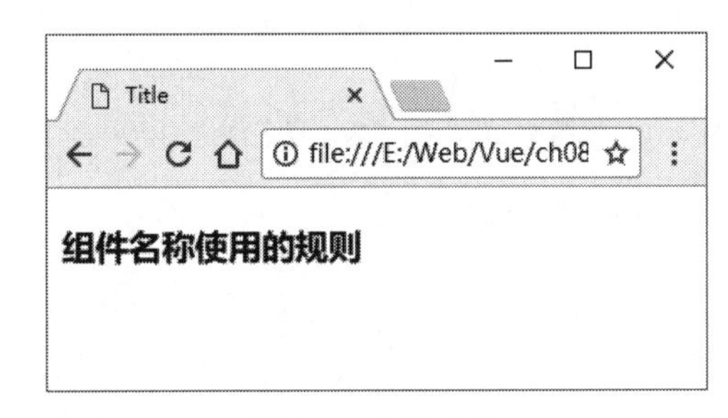

图8-2　组件命名

当然，也可以直接在Vue.component中以匿名对象的方式直接注册全局组件。

【例8.3】以匿名对象的方式注册组件。

```
<div id="app">
    <my-com2></my-com2><br/>
    <my-com3></my-com3>
</div>
<script>
    Vue.component('myCom2', Vue.extend({
        template: '<div>这是直接使用 Vue.component 创建的组件myCom2</div>'
    }))
    Vue.component('myCom3', {
        template: '<div>这是直接使用 Vue.component 创建的组件myCom3</div>'
    })
    var app = new Vue({
        el: '#app'
    });
</script>
```

在谷歌浏览器里面运行，效果如图8-3所示。

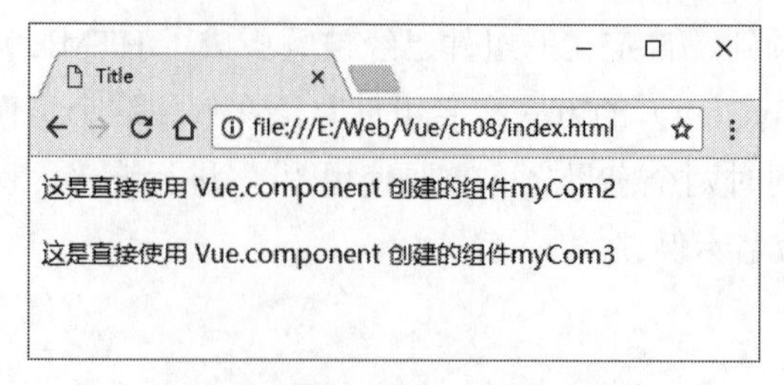

图 8-3　以匿名对象的方式注册组件

上面的示例中，只是在 template 属性中输入了一个简单的 HTML 代码，在实际的使用中，template 属性指向的模板内容可能包含很多的元素，而使用 Vue.extend 创建的模板必须有且只有一个根元素，出现多个根元素时，默认只渲染第一个根元素的内容。因此，当需要创建具有复杂元素的模板时，可以在最外层再套一个 div。例如，下面代码就是错误的编写方式。

【例 8.4】错误的编写方式。

```
<div id="app">
    <my-com></my-com>
</div>
<script>
    var comElement = Vue.extend({
        template: '<h3>全局组件</h3><p>这是我们创建的全局组件</p>'
    })
    Vue.component('my-com', comElement)
    var app = new Vue({
        el: '#app'
    });
</script>
```

在谷歌浏览器里面运行，可以发现只渲染了 <h3> 标签，<p> 标签没有被渲染，如图 8-4 所示。

图 8-4　错误的编写方式

只需要在“<h3>全局组件</h3><p>这是我们创建的全局组件</p>”外添加一个div即可：

```
<div><h3>全局组件</h3><p>这是我们创建的全局组件</p></div>
```

当template属性中包含很多的元素时，不能使用代码提示还是会显得不方便，这时，可以使用template标签来定义模板，通过id来确定组件的模板信息。

【例8.5】定义模板。

```
<div id="app">
    <my-com></my-com>
</div>
<template id="tmp">
    <div>
        <h3>Vue.js</h3>
        <h4>是现今最流行的框架之一</h4>
    </div>
</template>
<script>
    Vue.component('my-com',{
        template: '#tmp'
    })
    var vm = new Vue({
        el: '#app',
    });
</script>
```

在谷歌浏览器里面运行，效果如图8-5所示。

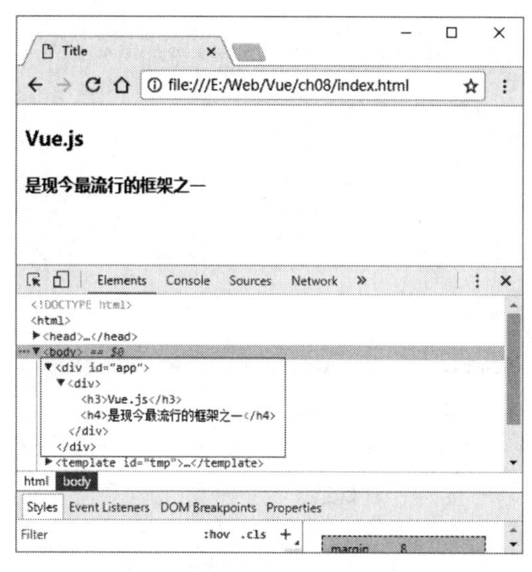

图8-5　定义模板

8.2.2　局部注册

有些时候，注册的组件只想在一个Vue实例中使用，如果还是使用全局注册的方式注册组件，就不太合适了。这时，可以使用局部注册的方式注册组件。

在 Vue 实例中，可以通过 components 属性注册仅在当前作用域下可用的组件。

【例 8.6】局部注册组件。

```
<div id="app1">
    <my-com></my-com>
</div>
<div id="app2">
    <my-com></my-com>
</div>
<template id="app2-com">
    <h4>app2中注册的局部组件</h4>
</template>
<script>
    var app1= new Vue({
        el: '#app1',
    });
    var app2 = new Vue({
        el: '#app2',
        components: {
            'my-com': {
                template: '#app2-com'
            }
        }
    });
</script>
```

在谷歌浏览器中运行，效果如图 8-6 所示。

图 8-6　局部注册组件

可以看到，在上面的示例中，局部注册的组件只能在注册的 app2 实例中完成解析，在 app1 实例中引用这个组件时，是无法正确解析这个自定义的组件元素的。

8.3 组件中的 data 选项

当一个 Vue 实例被创建之后，实例中的 data 选项的属性值就与页面的视图做了一个“绑定”，当修改 data 中的属性值时，视图就会产生“响应”，同时，页面上使用到属性值的地方也会同步更新。那么，组件作为一个特殊的 Vue 实例，对于 data 选项的使用上是不是和在 Vue 实例中的使用方式相同呢？

首先，还是先创建一个全局组件，按照 Vue 实例中使用 data 选项的使用方法，将 data 选项添加到组件的定义中，同时，使用插值表达式在页面中显示出属性值。

【例 8.7】测试 data 选项。

```
<div id="app">
    <my-com></my-com>
</div>
<template id="tmp">
    <div>
        <h4>{{name}}</h4>
    </div>
</template>
<script>
    Vue.component('my-com', {
        template: '#tmp',
        data: {
            name: '马云'
        }
    });
    var app = new Vue({
        el: '#app'
    });
</script>
```

在 IE 11 浏览器里面运行，可以发现页面上并没有显示出数据。

其实，在创建组件实例中的 data 选项时，返回的应该是一个实例对象的方法。把上面示例中的 data 属性改为方法：

```
Vue.component('my-com', {
        template: '#tmp',
        data: function () {
            return {
                name: '马云'
            }
        }
    });
```

在谷歌浏览器中运行，效果如图 8-7 所示。

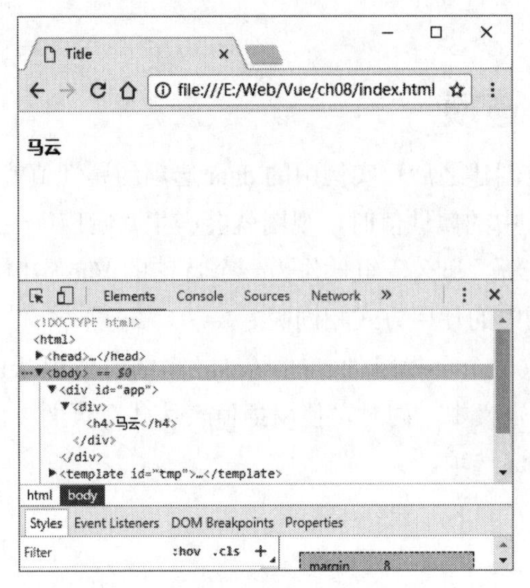

图 8-7　data 选项

为什么会这样呢？查看 Vue 的官方文档后，可以找到这样一句解释：一个组件的 data 选项必须是一个函数，因此每个实例可以维护一份被返回对象的独立的拷贝。

在 data 选项中将返回的对象改成在外部定义，同时多次调用这个组件。

【例 8.8】多次调用组件。

```
<div id="app">
    <counter></counter>
    <hr />
    <counter></counter>
</div>
<template id="counter">
    <div>
        <button @click="add"> Add </button>
        <h3>count: {{count}}</h3>
    </div>
</template>
<script>
    var dataObj = {
        count: 0
    }
    Vue.component('counter', {
        template: '#counter',
        data: function () {
            return dataObj
        },
        methods: {
            add() {
                dataObj.count++
            }
        },
    })
    var app= new Vue({
        el: '#app',
    });
</script>
```

在谷歌浏览器中运行，不论单击上面还是下面的Add按钮，count的值都会增加，效果如图8-8所示。

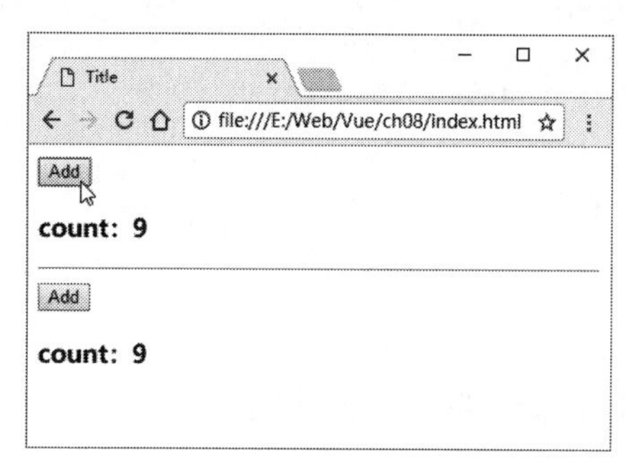

图8-8 多次调用组件效果

8.4 组件中的props选项

组件的props选项是在组件上注册的一些自定义特性，将一个值传递给一个props选项中的特性时，那个props特性就变成了组件实例的一个属性，这时就可以获取到这个值了。因此，组件中的props经常用于将父组件的值传递到子组件或是将Vue实例中的属性值传递到组件中使用。

在父组件引用子组件的时候，通过属性绑定的方式（v-bind），将需要传递给子组件的数据进行传递，从而在子组件内部，通过绑定的属性值获取到父组件的数据。

例如，在下面的代码中，在Vue实例中定义了一个局部组件，将实例的title属性绑定到组件的parenttitle属性上，同时将parenttitle属性赋值给组件的content属性。

【例8.9】测试父组件引用子组件。

```
<div id="app">
    <h4>
        请输入需要传递给子组件的值：<input type="text" v-model="title" />
    </h4>
    <child-node v-bind:parenttitle="title"></child-node>
</div>
<template id="child">
    <div>
        <h4>Vue 实例中的属性值为：{{content}}</h4>
    </div>
</template>
<script>
    var vm = new Vue({
        el: '#app',
        data: {
            title: ''
        },
        components: {
```

```
                'childNode': {
                    template: '#child',
                    props: ['parenttitle'],
                    data() {
                        return {
                            content: this.parenttitle
                        }
                    }
                }
            },
        });
</script>
```

在谷歌浏览器中运行，在输入框中输入“123456”，效果如图 8-9 所示。

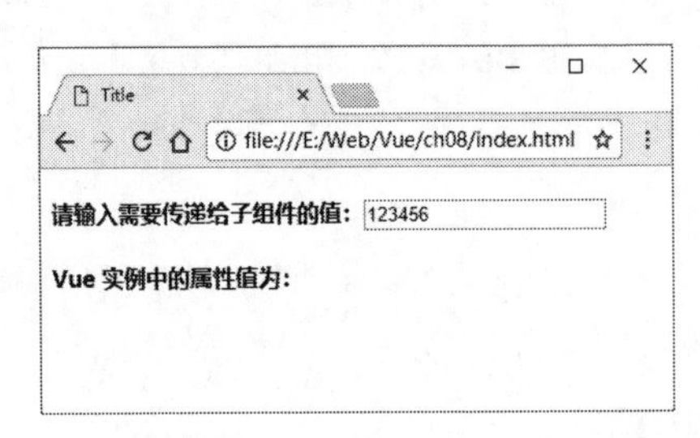

图 8-9　测试父组件引用子组件

可以看到，虽然可以在绑定的 parenttitle 属性上实时获取到 Vue 实例的属性值，可是却无法同步更新组件的 content 属性。组件的 data 选项中的 content 属性是一个 String 类型的变量，也就是 JavaScript 中的基本数据类型，在创建时就已经将数据值写入到内存栈中，之后与初始赋值的数据就没有任何的关系，因此后面对于 Vue 实例的 title 属性的任何变更，都不会引起对于实例 content 属性的变化。

因此，如果想完成最开始的需求的话，有两种可行的解决办法：

（1）将 Vue 实例中的 title 属性改为一个对象，输入的值作为对象中的一个属性，在实例中绑定的 parenttitle 也将作为一个对象。因为 JavaScript 中的对象为引用类型，赋值时，是将存储数据的地址进行拷贝，因此 title 对象和 parenttitle 对象将指向同一地址，任何对于 title 对象的属性值的修改，都会引起 parenttitle 对象中属性值的改变。

（2）采取 watch 监听 parenttitle 的方式来同步更新实例的 content 属性。

这里采用第二种方式，使用 watch 监听器来同步变更组件中的 content 属性值，修改代码如下。

【例 8.10】 watch 监听。

```
<script>
    var vm = new Vue({
        el: '#app',
        data: {
            title: ''
        },
        components: {
            'childNode': {
                template: '#child',
```

```
                props: ['parenttitle'],
                data() {
                    return {
                        content: this.parenttitle
                    }
                },
                watch: {
                    parenttitle:function() {
                        this.content = this.parenttitle
                    }
                },
            }
        },
    });
</script>
```

在谷歌浏览器中运行，在输入框中输入“123456”，可以发现 Vue 实例中的属性值随着输入框的内容而变化，效果如图 8-10 所示。

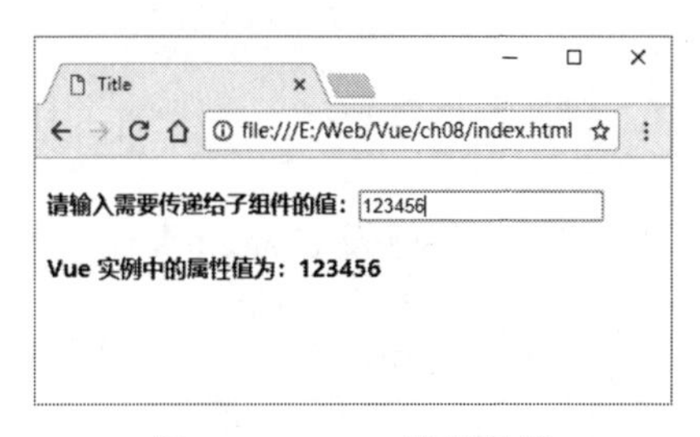

图 8-10　watch 监听效果

既然可以将 Vue 实例的数据传递到子组件中进行使用，那么，是不是可以在子组件进行修改绑定的属性值，从而影响到 Vue 实例呢？

在上面的实例代码的基础上，为子组件的 h4 标签添加一个单击事件 change，通过 change 事件来修改绑定的属性值 parenttitle，看一下会不会对 Vue 实例中的 title 属性造成影响。

【例 8.11】子组件修改绑定的属性值。

```
<div id="app">
    <h4>
        请输入需要传递给子组件的值：<input type="text" v-model="title" />
    </h4>
    <child-node v-bind:parenttitle="title"></child-node>
</div>
<template id="child">
    <div>
        <h4 @click="change">Vue 实例中的属性值为：{{content}}</h4>
    </div>
</template>
<script>
    var vm = new Vue({
        el: '#app',
        data: {
            title: ''
        },
```

```
            components: {
                'childNode': {
                    template: '#child',
                    props: ['parenttitle'],
                    data:function() {
                        return {
                            content: this.parenttitle
                        }
                    },
                    watch: {
                        parenttitle:function() {
                            this.content = this.parenttitle
                        }
                    },
                    methods: {
                        change::function() {
                            this.parenttitle = '123456'
                        }
                    }
                }
            },
        });
    </script>
```

在谷歌浏览器中运行，单击 h4 标签触发 change 事件，效果如图 8-11 所示。

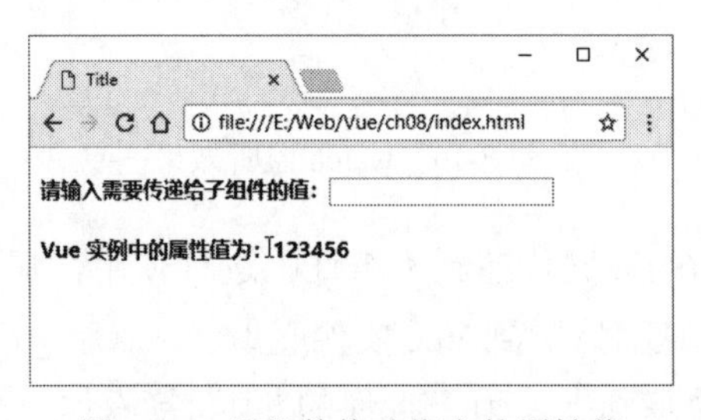

图 8-11　子组件修改绑定的属性值

可以看到，虽然修改了组件的 parenttitle 属性值，却没有影响到 Vue 实例中的 title 属性，所以可以得出，所有的 props 都使得其父子 props 之间形成了一个单向下行绑定：父级 props 的更新会向下流动到子组件中，但是反过来则不行。这样会防止从子组件意外改变父级组件的状态，从而导致应用的数据流向难以理解。

8.5 组件的复用

组件是可复用的 Vue 实例，且带有一个名字，在下面的示例中是 <button-counter>。我们可以在一个通过 new Vue 创建的 Vue 根实例中，把这个组件作为自定义元素来使用。

【例 8.12】创建 Vue 组件。

```
<div id="app">
```

```
    <button-counter></button-counter>
</div>
<script>
    //定义一个名为 button-counter 的组件
    Vue.component('button-counter', {
        data: function () {
            return {
                count: 0
            }
        },
        template: '<button v-on:click="count++">你单击了{{ count }}次</button>'
    }),
    new Vue({
        el: '#app',
    })
</script>
```

在谷歌浏览器中运行，不断单击按钮，count 值不断增加，效果如图 8-12 所示。

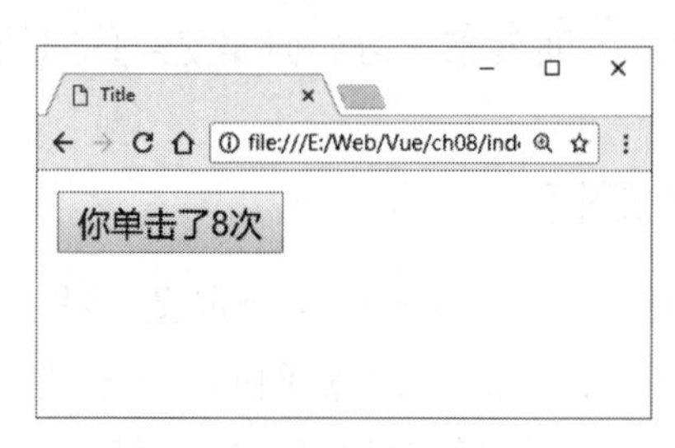

图 8-12　创建一个 Vue 组件

因为组件是可复用的 Vue 实例，所以它们与 new Vue 接收相同的选项，例如 data、computed、watch、methods 以及生命周期钩子等。仅有的例外是像 el 这样根实例特有的选项。

可以将组件进行任意次数的复用：例如修改上面的示例，复制模板，代码如下：

```
<div id="app">
    <button-counter></button-counter>
    <button-counter></button-counter>
    <button-counter></button-counter>
</div>
```

在谷歌浏览器中运行，效果如图 8-13 所示。

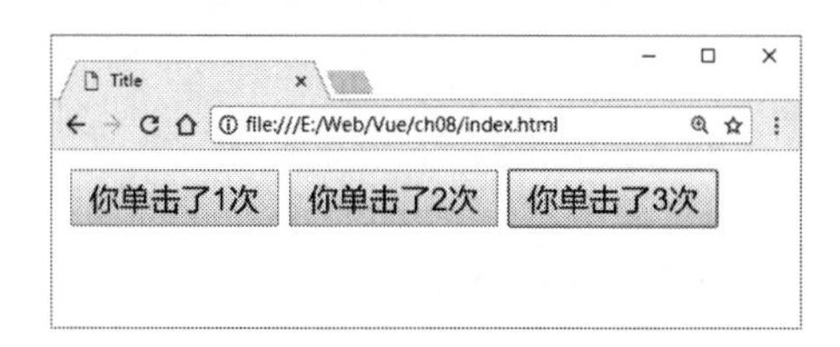

图 8-13　复用组件

提示

当单击按钮时，每个组件都会各自独立维护它的 count，因为每用一次组件，就会有一个它的新实例被创建。

这是因为如果像 Vue 实例那样，传入一个对象，由于 JavaScript 中对象类型的变量实际上保存的是对象的引用，所以当存在多个这样的组件时，会共享数据，导致一个组件中数据的改变会引起其他组件数据的改变。而使用一个返回对象的函数，每次使用组件都会创建一个新的对象，这样就不会出现共享数据的问题了。

8.6 组件间的数据通信

组件是 Vue 中的一个非常重要的概念，通过将页面拆分成一个个独立的组件，从而更好地实现代码复用，使代码更加简洁，容易维护。既然在 Vue 中会大量地使用到组件，必定会涉及组件之间的通信，下面就来学习如何实现组件间的数据交互。

8.6.1 父组件向子组件通信

在前面小节，提到使用 props 选项在组件中定义一些自定义特性，当有值传递给 props 特性时，props 特性就变成了那个组件实例的一个属性，此时就可以从组件上获取到接收的值。这里就可以在使用子组件时通过 v-bind 指令动态地绑定一个 props 特性，从而接收到父组件传递的值。

在下面的示例代码中，在子组件中通过 v-bind 指令绑定了一个 prop 特性 parenttitle，用来接收父组件 data 选项中的 title 属性，之后通过 watch 监听属性监听绑定的 parenttitle 属性，从而同步更新子组件 data 选项中的 content 属性值。

【例 8.13】父组件向子组件传递数据。

```
<body>
<div id="app">
    <h4>
        请输入需要传递给子组件的值: <input type="text" v-model="title" />
    </h4>
    <child-node v-bind:parenttitle="title"></child-node>
</div>
</body>
<template id="child">
    <div>
        <h5>
            Vue实例中的属性值为: {{content}}
        </h5>
    </div>
</template>
<script>
    var app= new Vue({
    el: '#app',
```

```
            data: {
                title: ''
            },
            components: {
                'childNode': {
                    template: '#child',
                    props: ['parenttitle'],
                    data() {
                        return {
                            content: this.parenttitle
                        }
                    },
                    watch: {
                        parenttitle() {
                            this.content = this.parenttitle
                        }
                    },
                }
            }
        })
    </script>
    </body>
```

在谷歌浏览器中运行，在输入框中输入“123456”，可以发现 Vue 实例中的属性值随着输入框的内容而变化，效果如图 8-14 所示。

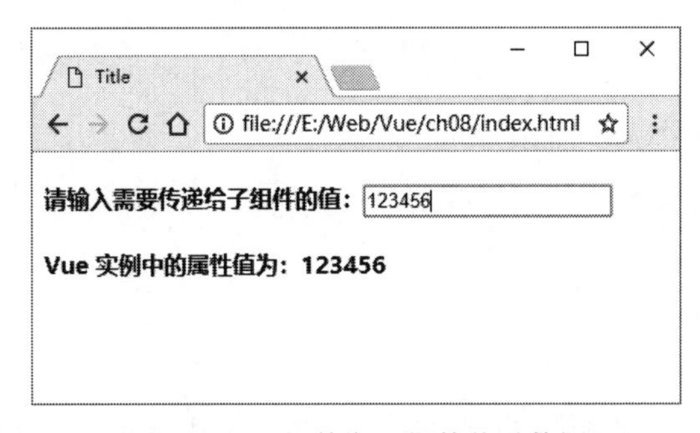

图 8-14　父组件向子组件传递数据

8.6.2　子组件向父组件通信

在 Vue 中存在着一个单向的下行绑定，父级组件的数据变更可以影响到子集组件，反过来则不行。在实际开发中可能会遇到当子组件的数据更新后，需要同步更新父组件的情况，那么这时应该怎么做呢？

想要实现目标，可以在子组件数据发生变化后，触发一个事件方法，告诉父组件数据更新了，父组件只需要监听这个事件，当捕获到这个事件运行后，再对父组件的数据进行同步变更，整个示意流程如图 8-15 所示。

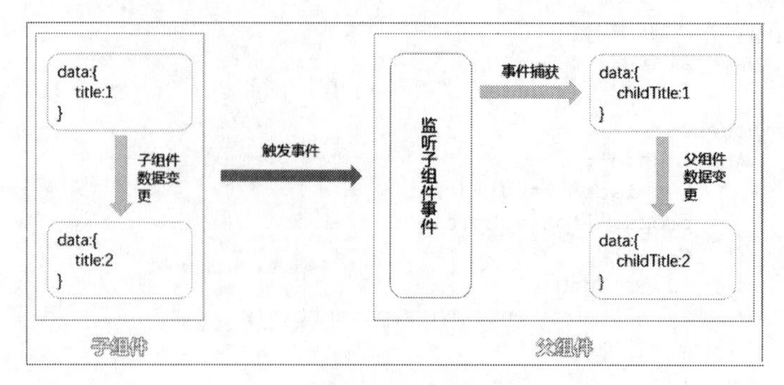

图 8-15 示意流程

可以使用 v-on 事件监听器监听事件，通过 $emit 去触发当前实例上的事件。当然，这里的事件可以是 JavaScript 中的原生 DOM 事件，也可以是自定义的事件。

在下面的代码中，当单击传递数据按钮后，触发了子组件的 func 方法，在 func 方法中触发子组件实例上的 show 事件，并把 input 框中的值作为参数进行传递。这时，在使用到子组件的地方就可以通过 v-on（@）指令监听这个 show 事件，从而获取到传递的参数，并触发父组件的监听事件。

【例 8.14】子组件向父组件传递数据。

```
<body>
<div id="app">
    <h5>
        子组件中的属性值为：{{msg}}
    </h5>
    <child-node @show="showMsg"></child-node>
</div>
</body>
<template id="child">
    <div>
        <h4>
            请输入需要传递给父组件的值：<input v-model="childMsg" type="text" />
            <button @click="func">传递数据</button>
        </h4>
    </div>
</template>
<script>
    var vm = new Vue({
        el: '#app',
        data: {
            title: '',
            msg: '',
        },
        methods: {
            showMsg(data) {
                this.msg = data
            }
        },
        components: {
            'childNode': {
                template: '#child',
                data() {
                    return {
                        childMsg: ''
```

```
                }
            },
            methods: {
                func() {
                    this.$emit('show', this.childMsg)
                }
            },
        }
    }
})
</script>
```

在谷歌浏览器中运行，在输入框中输入“123456”，然后单击“传递数据”按钮，效果如图 8-16 所示。

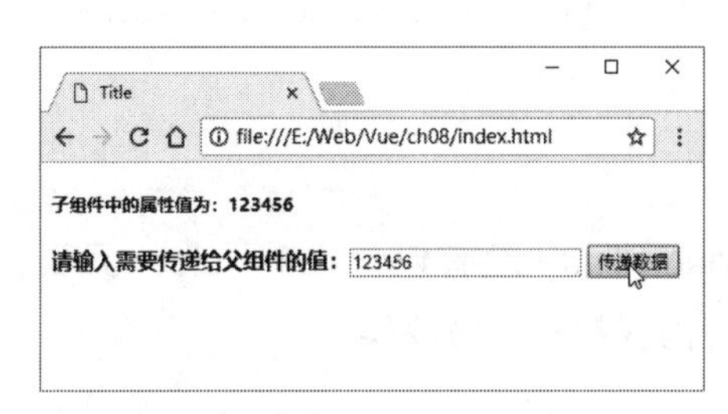

图 8-16　子组件向父组件传递数据

8.7 插槽

插槽，也就是 slot，它是组件的一块 HTML 模板，这块模板显示不显示以及怎样显示由父组件来决定。

8.7.1　认识插槽

下面从三方面来认识一下插槽：插槽内容、编译作用域和默认内容。

1. 插槽内容

Vue 实现了一套内容分发的 API，这套 API 的设计灵感源自 Web Components 规范草案，将 <slot> 元素作为承载分发内容的出口。

它允许合成组件，代码如下：

```
<navigation-link url="/profile">
  你的个人信息
</navigation-link>
```

然后在 <navigation-link> 的模板中编写以下代码：

```
<a v-bind:href="url" class="nav-link">
  <slot></slot>
</a>
```

当组件渲染的时候，<slot></slot> 将会被替换为“你的个人信息”。插槽内可以包含任何模板代码，包括 HTML：

```
<navigation-link url="/profile">
  <!--添加一个 Font Awesome图标 -->
  <span class="fa fa-user"></span>
  你的个人信息
</navigation-link>
```

插槽内甚至可以包含其他组件，代码如下：

```
<navigation-link url="/profile">
  <!-- 添加一个图标的组件 -->
  <font-awesome-icon name="user"></font-awesome-icon>
  你的个人信息
</navigation-link>
```

提示

如果 <navigation-link> 没有包含一个 <slot> 元素，则该组件起始标签和结束标签之间的任何内容都不会被渲染。

2. 编译作用域

当想在一个插槽中使用数据时，例如：

```
<navigation-link url="/profile">
  登录{{ user.name }}
</navigation-link>
```

该插槽跟模板的其他地方一样可以访问相同的实例属性（也就是相同的“作用域”），而不能访问 <navigation-link> 的作用域。例如 url 是访问不到的：

```
<navigation-link url="/profile">
  Clicking here will send you to: {{url}}
  <!--
   这里的 'url' 会是undefined，因为 "/profile"是_传递给_ <navigation-link> 的而不是在
<navigation-link> 组件*内部*定义的。
  -->
</navigation-link>
```

父级模板里的所有内容都是在父级作用域中编译的；子模板里的所有内容都是在子作用域中编译的。

3. 默认内容

有时为一个插槽设置具体默认的内容是很有用的，它只会在没有提供内容的时候被渲染。例如，<submit-button> 组件代码如下：

```
<button type="submit">
  <slot></slot>
</button>
```

如果希望这个 <button> 内绝大多数情况下都渲染文本“Submit”，为了将“Submit”作为默认内容，可以将它放在 <slot> 标签内：

```
<button type="submit">
  <slot>Submit</slot>
</button>
```

当在一个父级组件中使用 <submit-button> 并且不提供任何插槽内容时：

```
<submit-button></submit-button>
```

默认内容“Submit”将会被渲染，代码如下。

```
<button type="submit">
  Submit
</button>
```

【例 8.15】使用插槽示例。

```
<body>
<div id="app">
    <navigation-link url="/profile"></navigation-link>
</div>
<script>
        //注册navigation-link组件
        Vue.component('navigation-link', Vue.extend({
                    template:  '<div><button  type="submit"><slot>Submit</slot></
button></div>'
        }));
    var app= new Vue({
        el: '#app'
    });
</script>
</body>
```

在谷歌浏览器中运行，效果如图 8-17 所示。

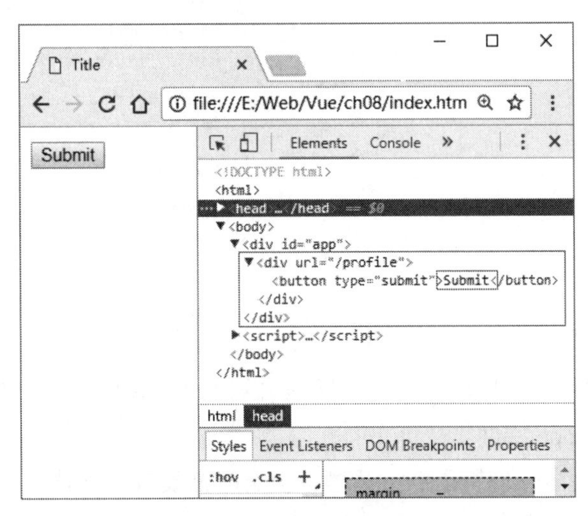

图 8-17　使用插槽

在父级组件中提供插槽内容“Save”时，这个提供的内容将会被渲染，从而取代默认的内容，代码如下。

```
<submit-button>
  Save
</submit-button>
```

【例 8.16】父级组件中提供插槽内容。

```
<body>
<div id="app">
    <navigation-link url="/profile">Save</navigation-link>
</div>
<script>
        //注册navigation-link组件
        Vue.component('navigation-link', Vue.extend({
          template: '<div><button type="submit"><slot>Submit</slot></button></div>'
        }));
    var app= new Vue({
        el: '#app'
    });
</script>
</body>
```

在谷歌浏览器中运行，效果如图 8-18 所示。

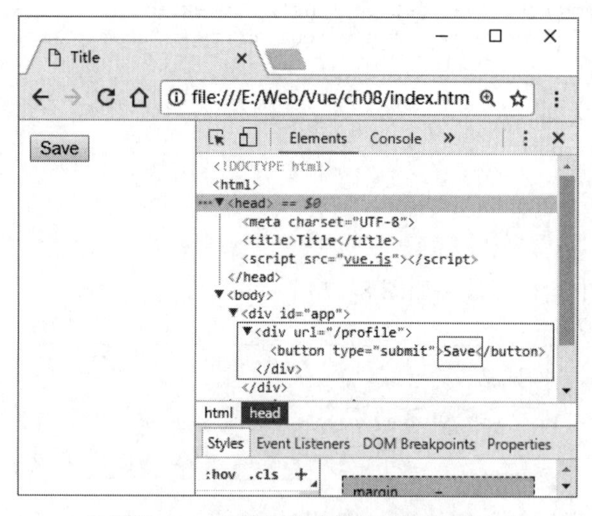

图 8-18　父级组件中提供插槽内容

8.7.2　具名插槽

在 Vue 2.6.0 中，具名插槽和作用域插槽引入了一个新的统一语法的 v-slot 指令。它取代了 slot 和 slot-scope 特性。

在上一节的示例中只使用了一个插槽，但有时需要多个插槽，例如对于一个带有如下模板的 <base-layout> 组件：

```
<div class="container">
  <header>
    <!-- 我们希望把页头放这里 -->
  </header>
  <main>
    <!-- 我们希望把主要内容放这里 -->
  </main>
  <footer>
    <!-- 我们希望把页脚放这里 -->
  </footer>
</div>
```

对于这样的情况，<slot> 元素有一个特性：name。这个特性可以用来定义额外的插槽：

```
<div class="container">
  <header>
    <slot name="header"></slot>
  </header>
  <main>
    <slot></slot>
  </main>
  <footer>
    <slot name="footer"></slot>
  </footer>
</div>
```

一个不带 name 的 <slot> 出口会带有隐含的名字 default。

在向具名插槽提供内容的时候，可以在一个 <template> 元素上使用 v-slot 指令，并以 v-slot 的参数的形式提供其名称：

```
<base-layout>
  <template v-slot:header>
    <h1>这里有一个页面标题</h1>
  </template>
  <p>这里有一段主要内容</p>
  <p>和另一个主要内容</p>
  <template v-slot:footer>
    <p>这是一些联系方式</p>
  </template>
</base-layout>
```

现在 <template> 元素中的所有内容都将会被传入相应的插槽。任何没有被包裹在带有 v-slot 的 <template> 中的内容都会被视为默认插槽的内容。

如果希望更明确一些，仍然可以在一个<template>中包裹默认插槽的内容，代码如下：

```
<base-layout>
  <template v-slot:header>
    <h1>这里有一个页面标题</h1>
  </template>
  <template v-slot:default>
    <p>这里有一段主要内容</p>
    <p>和另一个主要内容</p>
```

```
    </template>
    <template v-slot:footer>
        <<p>这是一些联系方式</p>
    </template>
</base-layout>
```

上面两种写法都会渲染出如下代码：

```
<div class="container">
    <header>
        <h3>这里有一个页面标题</h3>
    </header>
    <main>
        <p>这里有一段主要内容</p>
        <p>和另一个主要内容</p>
    </main>
    <footer>
        <p>这是一些联系方式</p>
    </footer>
</div>
```

【例 8.17】具名插槽示例。

```
<div id="app">
    <base-layout>
        <template v-slot:header>
            <h3>这里有一个页面标题</h3>
        </template>
        <template>
            <p>这里有一段主要内容</p>
            <p>和另一个主要内容</p>
        </template>
        <template v-slot:footer>
            <p>这是一些联系方式</p>
        </template>
    </base-layout>
</div>
<script>
        //注册base-layout组件
        Vue.component('base-layout', Vue.extend({
            template: '<div class="container">\n' +
                '<header><slot name="header"></slot></header>\n' +
                '<main><slot></slot></main>\n' +
                '<footer><slot name="footer"></slot></footer>\n' +
                '</div>'
        }));
    var app= new Vue({
        el: '#app'
    });
</script>
</body>
```

在谷歌浏览器中运行，效果如图 8-19 所示。

图 8-19　具名插槽示例

跟 v-on 和 v-bind 一样，v-slot 也有缩写，即把参数之前的所有内容 (v-slot:) 替换为字符 #。例如 v-slot:header 可以被重写为 #header：

```
<base-layout>
        <template #header>
            <h1>这里有一个页面标题</h1>
        </template>
        <template #default>
            <p>这里有一段主要内容</p>
            <p>和另一个主要内容</p>
        </template>
        <template #footer>
            <<p>这是一些联系方式</p>
        </template>
    </base-layout>
```

8.7.3　作用域插槽

提示

自 Vue 2.6.0 已废弃使用 slot-scope 特性。

有时让插槽内容能够访问子组件中才有的数据是很有用的。例如，设想一个带有如下模板的 <current-user> 组件：

```
<span>
  <slot>{{ user.lastName }}</slot>
</span>
```

想让它的默认内容显示用户的名，以取代正常情况下用户的姓，编写如下代码：

```
<current-user>
  {{ user.firstName }}
</current-user>
```

然而上述代码不会正常工作，因为只有<current-user>组件可以访问到user，而提供的内容是在父级渲染的。

为了让user在父级的插槽内容可用，可以将user作为<slot>元素的一个特性绑定上去：

```
<span>
  <slot v-bind:user="user">
    {{ user.lastName }}
  </slot>
</span>
```

绑定在<slot>元素上的特性被称为插槽prop。现在，在父级作用域中，可以给v-slot带一个值，来定义提供的插槽prop的名字：

```
<current-user>
  <template v-slot:default="slotProps">
    {{ slotProps.user.firstName }}
  </template>
</current-user>
```

在这个例子中，选择将包含所有插槽prop的对象命名为slotProps，也可以使用任意名字，具体的示例代码如下。

【例 8.18】作用域插槽示例。

```
<div id="app">
    <current-user>
            <template v-slot:default="slotProps">
                {{slotProps.user.firstName }}
            </template>
    </current-user>
</div>
<script>
    Vue.component('currentUser', {
            template: ' <span><slot :user="user">{{ user.lastName }}</slot></
span>',
        data:function(){
            return {
                user: {
                    firstName: '启灵',
                    lastName: '张'
                }
            }
        }
    });
    new Vue({
        el: '#app'
    })
</script>
</body>
```

在谷歌浏览器中运行，效果如图 8-20 所示。

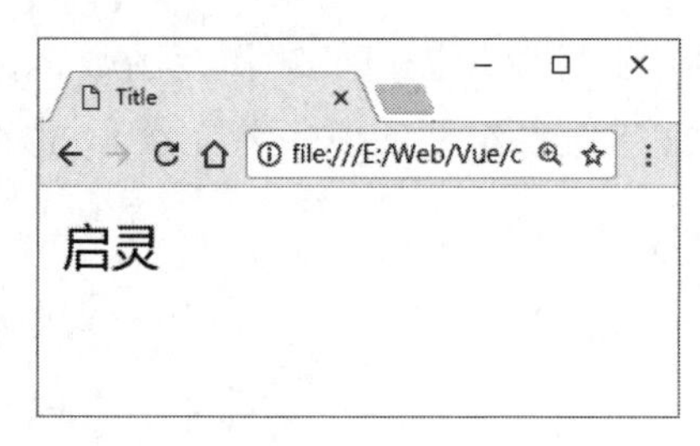

图 8-20　作用域插槽

在上述情况下，当被提供的内容只有默认插槽时，组件的标签才可以被当作插槽的模板来使用。这样就可以把 v-slot 直接用在组件上：

```
<current-user v-slot:default="slotProps">
  {{ slotProps.user.firstName }}
</current-user>
```

这种写法还可以更简单，就像假定未指明的内容对应默认插槽一样，不带参数的 v-slot 被假定对应默认插槽：

```
<current-user v-slot="slotProps">
  {{ slotProps.user.firstName }}
</current-user>
```

注意，默认插槽的缩写语法不能和具名插槽混用，因为它会导致作用域不明确：

```
<!-- 无效，会导致警告 -->
<current-user v-slot="slotProps">
  {{ slotProps.user.firstName }}
  <template v-slot:other="otherSlotProps">
    slotProps is NOT available here
  </template>
</current-user>
```

只要出现多个插槽，应始终为所有的插槽使用完整的基于 <template> 的语法：

```
<current-user>
  <template v-slot:default="slotProps">
    {{ slotProps.user.firstName }}
  </template>
  <template v-slot:other="otherSlotProps">
    ...
  </template>
</current-user>
```

8.7.4　解构插槽

作用域插槽的内部工作原理是将插槽内容包括在一个传入单个参数的函数里：

```
function (slotProps) {
  //插槽内容
}
```

这意味着 v-slot 的值实际上可以是任何能够作为函数定义中的参数的 JavaScript 表达式。所以在支持的环境下 (单文件组件或现代浏览器)，也可以使用 ES2015 解构来传入具体的插槽 prop，如下：

```
<current-user v-slot="{ user }">
  {{ user.firstName }}
</current-user>
```

这样可以使模板更简洁，尤其是在该插槽提供了多个 prop 的时候。它同样开启了 prop 重命名等其他可能，例如将 user 重命名为 person：

```
<current-user v-slot="{ user: person }">
  {{ person.firstName }}
</current-user>
```

甚至可以定义默认的内容，用于插槽 prop 是 undefined 的情形：

```
<current-user v-slot="{ user = { firstName: 'Guest' } }">
  {{ user.firstName }}
</current-user>
```

【例 8.19】解构插槽示例。

```
<body>
<div id="app">
    <current-user>
        <template v-slot="{user:person}">
            {{person.firstName }}
        </template>
    </current-user>
</div>
<script>
    Vue.component('currentUser', {
        template: ' <span><slot :user="user">{{ user.lastName }}</slot></span>',
        data:function(){
            return {
                user: {
                    firstName: '启灵',
                    lastName: '张'
                }
            }
        }
    });
    new Vue({
        el: '#app'
    })
</script>
</body>
```

在谷歌浏览器中运行，效果如图 8-21 所示。

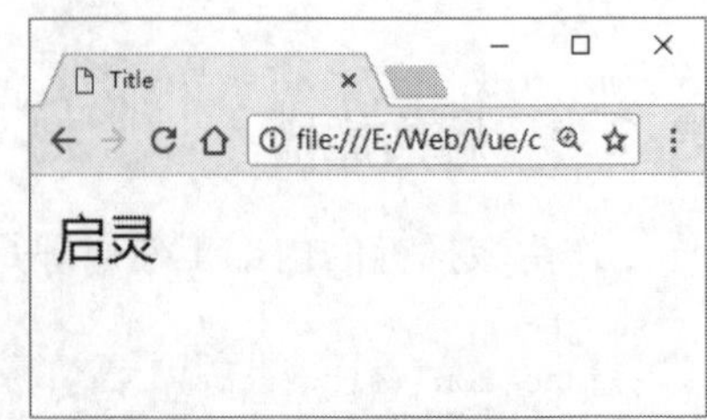

图 8-21　解构插槽示例

8.8 案例实战——设计照片相册

本案例是一个简单的相册展示页面，类似于轮播图效果，主要实现的功能如下。

（1）可使用左右图标箭头切换图片。

（2）可使用键盘上的左右方向键切换图片。

（3）可通过下面的缩略图选择指定图片。

在 HTML 页面中引入 vue-gallery 组件，主要内容都是在该组件中完成。代码如下：

```
<!DOCTYPE html>
<html>
<head>
<meta charset="UTF-8">
<title> Vue.js构建的简单照片相册</title>
<meta name="viewport" content="width=device-width, initial-scale=1">
<!--图标库-->
<link rel='stylesheet' href='https://cdnjs.cloudflare.com/ajax/libs/font-
   awesome/5.8.2/css/all.min.css'>
<!--核心样式-->
<link rel="stylesheet" href="style.css">
<!--引入vue.js框架-->
<script src='vue.min.js'></script>
</head>
<body>
<div class="header">
<h1>简单的照片相册</h1>
</div>
<div class="container" id="app">
<vue-gallery :photos="photos"></vue-gallery>
</div>
</body>
</html>
```

在 script 标签中定义了组件 vue-gallery，所有的内容都是在该组件中实现。其中，使用 props 属性传递图片的 URL 数据。具体实现代码如下：

```
<script>
//定义组件vue-gallery
Vue.component('vue-gallery', {
  props: ['photos'],
     data: function () {
          return {
              activePhoto: null
          }
     },
     //定义模板，在模板中定义切换按钮，图标使用font-awesome字体库
     template:'
    <div class="gallery">
    <div class="activePhoto" :style="'background-image: url('+photos[activePho
       to]+');'">
      <button type="button" class="previous" @click="previousPhoto()">
        <i class="fas fa-chevron-circle-left"></i>
      </button>
```

```
      <button type="button" class="next" @click="nextPhoto()">
        <i class="fas fa-chevron-circle-right"></i>
      </button>
    </div>
    <div class="thumbnails">
      <div
           v-for="(photo, index) in photos"
           :src="photo"
           :key="index"
           @click="changePhoto(index)"
           :class="{'active': activePhoto == index}" :style="'background-image:
              url('+photo+')'">
      </div>
    </div>
  </div>',
     mounted () {
           this.changePhoto(0);
           //document.addEventListener()方法用于向文档添加事件句柄
           document.addEventListener( "keydown", (event) => {
               //定义键盘上左右方向键控制
               if (event.which == 37)
                  this.previousPhoto()
               if (event.which == 39)
                  this.nextPhoto()
           })
     },
     methods: {
           changePhoto (index) {
               this.activePhoto = index
           },
           nextPhoto () {
               this.changePhoto( this.activePhoto+1 < this.photos.length ?
                  this.activePhoto+1 : 0 )
           },
           previousPhoto () {
               this.changePhoto( this.activePhoto-1 >=0 ? this.activePhoto-1 :
                  this.photos.length-1 )
           }
     }
  });
  new Vue({
     el: '#app',
     data: {
           photos:[
               'img/001.png',
               'img/002.png',
               'img/003.png',
               'img/004.png',
               'img/005.png',
               'img/006.png',
               'img/007.png',
               'img/008.png'
           ]
     }
  });
</script>
```

相应的样式代码如下：

```
body {
  background-color: #5c4090;
  padding: 25px;
}
.header {
  text-align: center;
}
#app{
  margin: auto;
}
.header h1 {
  background:-webkit-linear-gradient(#fff,#8f70ba);
  -webkit-text-fill-color: transparent;
  -webkit-background-clip: text;
  text-align: center;
  font-weight: 900;
  font-size: 3rem;
  color: #fff;
  margin-bottom: 30px;
}
.container {
  padding: 6px;
  background-color: #fff;
  border-radius: 8px;
  max-width: 800px;
  box-shadow: 0 5px 8px #0000007a;
}
.gallery .activePhoto {
  width: 100%;
  margin-bottom: 5px;
  padding-bottom: 65%;
  background-size: cover;
  background-position: center;
  background-repeat: no-repeat;
  border: 2px solid #fff;
  position: relative;
}
.gallery .activePhoto button {
  border: none;
  background-color: transparent;
  font-size: 32px;
  color: #fff;
  opacity: 0.5;
  position: absolute;
  outline: none;
  height: 100%;
}
.gallery .activePhoto button:hover {
  opacity: 1;
}
.gallery .activePhoto button.previous {
  padding: 0 1em 0 0.7em;
  left: 0;
  background: linear-gradient(to right, rgba(0, 0, 0, 0.5) 0%, rgba(0, 0, 0, 0)
    100%);
}
```

```
.gallery .activePhoto button.next{
  padding: 0 0.7em 0 1em;
  right: 0;
  background: linear-gradient(to right, rgba(0, 0, 0, 0) 0%, rgba(0, 0, 0, 0.5)
100%);
}
.gallery .thumbnails {
  display: grid;
  grid-template-columns: repeat(auto-fill, minmax(140px,1fr));
  grid-gap: 5px;
}
.gallery .thumbnails div {
  width: 100%;
  border: 2px solid #fff;
  outline: 2px solid #fff;
  cursor: pointer;
  padding-bottom: 65%;
  background-size: cover;
  background-position: center;
  background-repeat: no-repeat;
  opacity: 1;
}
.gallery .thumbnails div:hover{
  opacity: 0.6;
}
.gallery .thumbnails div.active{
  outline-color: #5c4084;
  opacity: 1;
}
```

在谷歌浏览器中运行，效果如图 8-22 所示。

图 8-22　相册效果

8.9 疑难解惑

疑问 1：对于刚学习 Vue 的读者，在写组件时，可能会写成如下代码：

```
<template>
    <h3>我是组件</h3>
    <strong>我是加粗标签</strong>
</template>
```

在谷歌浏览器中运行报如下错误：

```
▸ [Vue warn]: Error compiling template:                                    vue.js:634

  Component template should contain exactly one root element. If you are using v-if on multiple
  elements, use v-else-if to chain them instead.
```

解决方法：在 vue 2.x 中，每个组件模板中不再支持片段代码。必须由根元素包裹住所有的代码。把上面代码改成如下代码即可。

```
<template>
    <div>
        <h3>我是组件</h3>
        <strong>我是加粗标签</strong>
    </div>
</template>
```

疑问 2：组件的 data 数据为什么必须以函数返回的形式而不是简单的对象形式呢？

组件是可复用的 vue 实例，一个组件被创建好之后，就可能被用在各个地方，而组件不管被复用了多少次，组件中的 data 数据都应该是相互隔离，互不影响的，基于这一理念，组件每复用一次，data 数据就应该被复制一次，之后，当某一处复用的地方组件内 data 数据被改变时，其他复用地方组件的 data 数据不受影响，例如下面这个例子：

```
<div id="app">
    <button-counter></button-counter>
    <button-counter></button-counter>
    <button-counter></button-counter>
</div>
<script>
    //定义一个名为 button-counter 的组件
    Vue.component('button-counter', {
        data: function () {
            return {
                count: 0
            }
        },
        template: '<button v-on:click="count++">你单击了{{ count }}次</button>'
    }),
        new Vue({
            el: '#app',
        })
</script>
```

在谷歌浏览器中运行，如图 8-23 所示。

图 8-23　运行效果

该组件被复用了三次，但每个复用的地方组件内的 count 数据相互不受影响，它们各自维护各自内部的 count。单击第一个组件，可发现其他组件并没有任何变化，如图 8-24 所示。

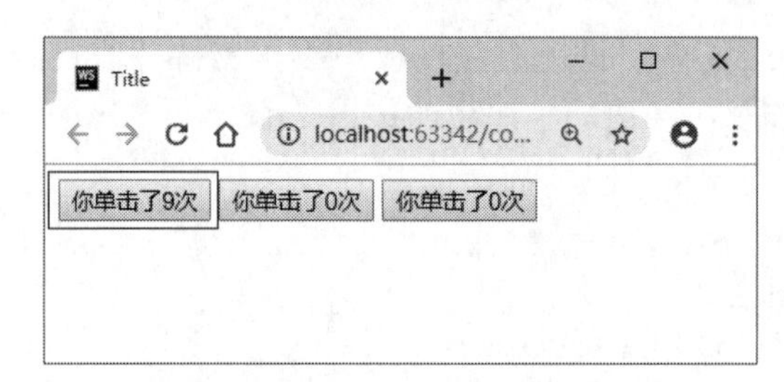

图 8-24　单击第一个组件按钮后的效果

能有这样的效果正是因为上述例子中的 data 不是一个单纯的对象，而是一个函数返回值的形式，所以每个组件实例可以维护一份被返回对象的独立拷贝。

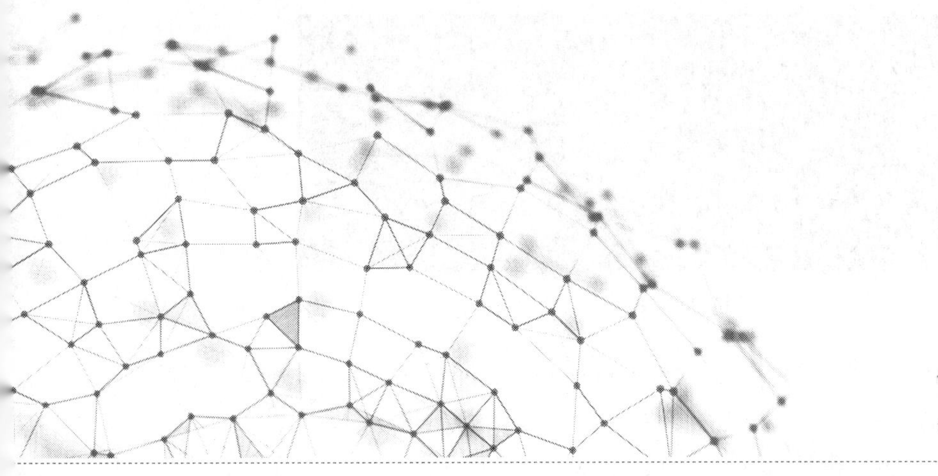

第9章

使用webpack打包

高效的开发离不开基础工程的搭建。本章主要介绍目前热门的 JavaScript 应用程序的模块打包工具 webpack。在开始学习本章前，需要先安装 Node.js 和 NPM，如果不熟悉它们，可以先查阅相关资料，完成安装并了解 NPM 最基本的用法。

9.1 前端工程化与 webpack

近几年来，前端领域发展迅速，前端的工作早已不再是切几张图那么简单，项目比较大时，可能会多人协同开发。模块化、组件化、CSS 预编译等概念也成了开发人员经常讨论的话题。

通常，前端自动化（半自动化）工程主要解决以下问题：

- JavaScript、CSS 代码的合并和压缩。
- CSS 预处理：Less、Sass、Stylus 的编译。
- 生成雪碧图（CSS Sprite）。
- ES 6 转 ES 5。
- 模块化。

如果使用过 Gulp，并且了解 RequireJS，那上面几个问题应该很容易解决。只需配置几行代码，就可以实现对 JS 代码的合并与压缩。不过，经过 Gulp 合并压缩后的代码仍然是用户自己写的代码，只是局部变量名被替换，一些语法做了转换而已，整体内容并没有发生变化。而本章要介绍的前端工程化工具 webpack，打包后的代码已经不只是用户自己写的代码，其中夹杂了很多 webpack 自身的模块处理代码。因此，学习 webpack 最难的是理解“编译”的这个概念。图 9-1 是来自 webpack 中文网站（https://www.webpackjs.com/）经典的模块化示意图。

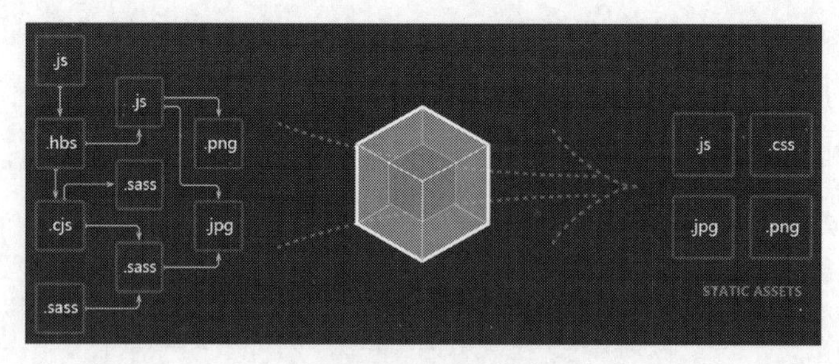

图 9-1　模块化示意图

左边是在业务中写的各种格式的文件，例如 typescript、less、jpg，还有本章后面要介绍的 .vue 格式的文件。这些格式的文件通过特定的加载器（Loader）编译后，最终统一生成为 .js、.css、.png 等静态资源文件。在 webpack 的世界里，一张图片、一个 CSS 甚至一个字体，都称为模块（Module），彼此存在依赖关系，webpack 就是来处理模块间的依赖关系的，并把它们进行打包。

举一个简单的例子，平时加载 CSS 大多通过 <link> 标签引入 CSS 文件，而在 webpack 里，直接在一个 .js 文件中导入，例如：

```
import ' src/styles/index. css ';
```

import 是 ES 2015 的语法，这里也可以写成 require('src/styles/index.css')。在打包时，index.css 会被打包进一个 .js 文件里，通过动态创建 <style> 的形式来加载 CSS 样式，当然也可以进一步配置，在打包编译时把所有的 CSS 都提取出来，生成一个 CSS 的文件，后面会详细介绍。

webpack 的主要适用场景是单页面富应用（SPA）。SPA 通常是由一个 HTML 文件和一堆按需加载的 JavaScript 组成，它的 HTLM 结构可能会非常简单，例如：

```
<!doctype html>
<html>
<head>
    <meta charset="UTF-8">
    <title></title>
    <link rel="stylesheet" href="dist/main.css">
</head>
<body>
<div id="app"></div>
<script src="dist/main.js"></script>
</body>
</html>
```

代码中只有一个 <div> 节点，所有的代码都集成到了 main.js 文件中，但它可以实现像知乎、淘宝这样大型的项目。

在开始讲解 webpack 的用法前，先介绍两个 ES 6 中的语法：export 和 import，因为在后面会大量使用，如果对它不了解，读者可能会感到很困惑。

export 和 import 是用来导出和导入模块的。一个模块就是一个 js 文件，它拥有独立的作用域，里面定义的变量外部是无法获取的。例如，将一个配置文件作为模块导出，示例代码如下：

```
//config.js
    var Config={
        version:'1.0.0'
    };
    export{Config};
```

或：

```
    //config.js
    export var Config={
        version:'1.0.0'
    };
```

其他类型（比如函数、数组、常量等）也可以导出，例如，导出一个函数：

```
//add.js
    export function add(a,b) {
        return a+b;
    }
```

模块导出后，在需要使用模块的文件使用 import 再导入，就可以在这个文件内使用这些模块了。示例代码如下：

```
//main.js
    import {Config} from './config.js';
    import {add} from './add.js';
    console.log(Config);    //{ version:'1.0.0'}
    console.log(add(1,1));  //2
```

以上几个示例中，导入的模块名称都是在 export 的文件中设置的，也就是说用户必须预先知道它的名称叫什么，例如 Config、add。而有的时候，用户不想去了解它的名称是什么，只是把模块的功能拿来使用，或者想自定义名称，这时可以使用 export default 来输出默认的模块。示例代码如下：

```
//config.js
    export default{
        version:'1.0.0'
    }
    //add.js
    export default function (a,b){
        return a+b;
    }
    //main.js
    import conf from './config.js';
    import Add from './add.js';
    console.log(conf)  //{version:'1.0.0'}
    console.log(Add(1,1))  //2
```

如果使用 NPM 安装了一些库，在 webpack 中可以直接导入，示例代码如下：

```
import Vue from 'vue';
import $ from 'jquery';
```

上例分别导入了 Vue 和 jQuery 的库，并且命名为 Vue 和 $，在这个文件中就可以使

用这两个模块。

export 和 import 还有其他的用法，这里不做太详细的介绍，读者如果有兴趣，可以查阅相关资料进一步学习。

9.2 webpack 基础配置

在使用 webpack 前，首先需要进行一些环境配置。

9.2.1 安装 webpack 与 webpack-dev-server

本节将从基础的 webpack 安装开始介绍，逐步完成对 Vue 工程的配置。在开始学习本节前，先确保已经安装了最新版的 Node.js 和 NPM，并已经了解 NPM 的基本用法。首先，创建一个目录，例如 demo，使用 NPM 初始化配置：

```
npm init
```

执行后，会有一系列选项，可以按 Enter 键快速确认，完成后会在 demo 目录生成一个 package.json 的文件。

之后在本地局部安装 webpack：

```
npm install webpack --save-dev
```

--save-dev 会作为开发依赖来安装 webpack。安装成功后，在 package.json 中会多一项配置：

```
"devDependencies": {
    "webpack": "^4.35.2"
  }
```

提示

npm install 在安装 npm 包时，有两种命令参数可以把它们的信息写入 package.json 文件，一个是 npm install --save，另一个是 npm install --save-dev，表面上的区别是 --save 会把依赖包名称添加到 package.json 文件的 dependencies 选项下，--save-dev 则添加到 package.json 文件的 devDependencies 选项下。它们真正的区别是，devDependencies 中的模块，是开发时用的，例如我们安装 js 的压缩包 gulp-uglify 时采用的是“npm install --save-dev gulp-uglify ”命令安装，因为在发布后用不到它，而只是在开发才用到它。Dependencies 中的模块，则是发布后还需要依赖的模块，例如像 jQuery 库或者 Angular 框架等，在开发完后还要依赖它们，否则就运行不了。

如果使用 webpack 4+ 版本，还需要安装脚手架（CLI）。

```
npm install --save-dev webpack-cli
```

通过以下的 NPM 安装方式，将使 webpack 在全局环境下可用：

```
npm install --global webpack
```

提示

不推荐全局安装 webpack。这会将项目中的 webpack 锁定到指定版本，在使用不同的 webpack 版本的项目中，可能会导致构建失败。

接着需要安装 webpack-dev-server，它可以在开发环境中提供很多服务，例如启动一个服务器、热更新、接口代理等，配置起来也很简单。在本地局部安装代码如下：

```
npm install webpack-dev-server --save-dev
```

安装完成后，最终的 package.json 文件内容如下：

```
{
  "name": "demo",
  "version": "1.0.0",
  "description": "",
  "main": "index.js",
  "scripts": {
    "test": "echo \"Error: no test specified\" && exit 1"
  },
  "author": "",
  "license": "ISC",
  "devDependencies": {
    "webpack": "^4.35.2",
    "webpack-dev-server": "^3.7.2"
  }
}
```

如果 devDependencies 中包含 webpack 和 webpack-dev-server，说明已经安装成功，可以启动 webpack 工程了。

9.2.2 webpack 的核心概念

接下来需要了解 webpack 的一些核心概念。

归根到底，webpack 就是一个 .js 配置文件，架构的好或差都体现在这个配置里，随着需求的不断出现，工程配置也是逐渐完善的。我们由浅入深，一步步来完成更多的功能。首先在目录 demo 下创建一个 js 文件 webpack.config.js，并初始化它的内容：

```
var config={
    //
};
module.exports=config;
```

提示

这里的 module.exports=config；相当于 export default config；。由于目前还没有安装支持 ES 6 的编译插件，因此不能直接使用 ES 6 的语法，否则会报错。

然后在 package.json 的 scripts 里增加一个快速启动 webpack-dev-server 服务的脚本：

```
"scripts": {
    "test": "echo \"Error: no test specified\" && exit 1",
    "dev":"webpack-dev-server --open --config webpack.config.js"
  },
```

当运行 npm run dev 命令时，就会执行 webpack-dev-server --open --config webpack.config.js 命令。其中 --config 是指向 webpack-dev-server 读取的配置文件路径，这里直接读取我们在上一步创建的 webpack.config.js 文件。--open 会在执行命令时自动在浏览器打开页面，默认地址是 127.0.0.1:8080，不过 IP 和端口都是可以配置的，例如：

```
"scripts": {
    "test": "echo \"Error: no test specified\" && exit 1",
     "dev":"webpack-dev-server --host 192.168.0.105 --port 8888 --open --config
webpack.config.js"
  },
```

这样访问地址就改为了 192.168.0.105:8888。一般在局域网下，需要让其他同事访问时可以这样配置，否则用默认的 127.0.0.1（localhost）就可以了。

webpack 配置中最重要也是必选的两项是入口（Entry）和出口（Output）。入口的作用是告诉 webpack 从哪里开始寻找依赖，并且编译，出口则用来配置编译后的文件存储位置和文件名。

在 demo 目录下新建一个空的 main.js 作为入口的文件，然后在 webpack.config.js 中进行入口和输出的配置：

```
var path=require('path');
var config={
    entry:{
        main:'./main'
    },
    output:{
        path:path.join(__dirname,'./dist'),
        publicPath:'/dist/',
        filename:'main.js'
    }
};
module.exports=config;
```

entry 中的 main 就是配置的单入口，webpack 会从 main.js 文件开始工作。output 中

path 选项用来存放打包后文件的输出目录，是必填项。publicPath 指定资源文件引用的目录，如果资源存放在 CDN 上，这里可以填 CDN 的网址。filename 用于指定输出文件的名称。因此，这里配置的 output 意为打包后的文件会存储为 demo/dist/main.js 文件，只要在 HTML 中引入它就可以了。

在 demo 目录下，新建一个 index.html 作为 SPA 的入口：

```
<!doctype html>
<html>
<head>
    <meta charset="UTF-8">
    <meta name="viewport"
              content="width=device-width, user-scalable=no, initial-scale=1.0,
maximum-scale=1.0, minimum-scale=1.0">
    <meta http-equiv="X-UA-Compatible" content="ie=edge">
    <title>Document</title>
</head>
<body>
<div id="app">
    <h1>hello world!</h1>
</div>
<script src="/dist/main.js"></script>
</body>
</html>
```

现在在终端执行下面的命令，就会自动在浏览器中打开页面了：

```
npm run dev
```

如果打开的页面和图 9-2 一致，那么已经完成整个工程中最重要的一步了。

图 9-2　页面效果

打开 demo/main.js 文件，添加一行 JavaScript 代码来修改页面的内容：

```
document.getElementById ('app') .innerHTML ='Hello webpack!';
```

保存文件，回到刚才打开的页面，发现页面内容已经变为“Hello webpack.”。注意，此时并没有刷新浏览器，就已经自动更新了，这就是 webpack-dev-server 的热更新功能，它通过建立一个 WebSocket 连接来实时响应代码的修改。

在 9.1 节中介绍过：学习 webpack 最难的是理解它“编译”的这个概念，在 Chrome 浏览器开发者工具的 network 视图中查看 main.js 的内容，如图 9-3 所示。

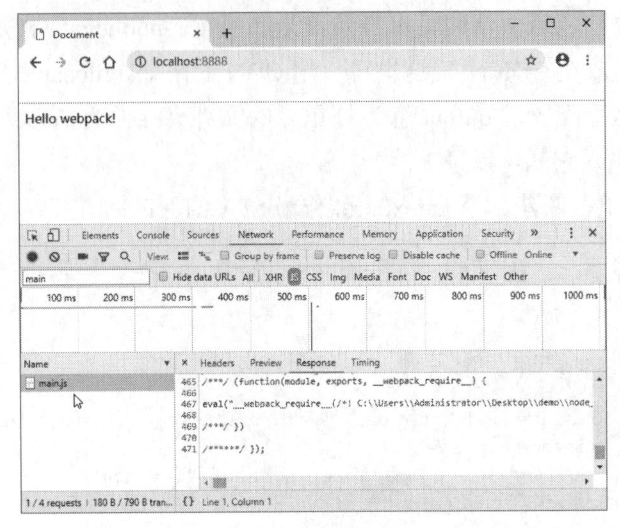

图 9-3　编译内容

我们只写了一行 JS 代码，却编译出了 471 行，虽然内容较多，但很多是 webpack-dev-server 的功能，只在开发时有效，在生产环境下编译就不会这么臃肿了。例如执行下面的命令进行打包：

```
webpack --progress --hide-modules
```

完成后，会生成一个 demo/dist/main.js 文件。

9.2.3　完善配置文件

通过配置入口（Entry）和出口（Output）已经可以启动 webpack 项目了，不过这并不是 webpack 的特点，如果它只有这些作用，根本就不用这么麻烦。本节将对文件 webpack.config.j 进一步配置，来实现更强大的功能。

在 webpack 的世界里，每个文件都是一个模块，比如 .css、.js、.html、.jpg、.less 等。对于不同的模块，需要用不同的加载器（Loaders）来处理，而加载器就是 webpack 最重要的功能。通过安装不同的加载器可以对各种后缀名的文件进行处理，比如现在要写一些 CSS 样式，就要用到 style-loader 和 css-loader。下面就通过 NPM 来安装它们：

```
npm install css-loader --save-dev
npm install style-loader --save-dev
```

安装完成后，在 webpack.config.js 文件里配置 Loader，增加对 .css 文件的处理：

```
var config={
    //…
    //css文件处理
    module:{
        rules:[
            {
                test:/\.css$/,
```

```
                    use:[
                        'style-loader',
                        'css-loader'
                    ]
                }
            ]
        }
};
```

在 module 对象的 rules 属性中可以指定一系列的 Loader，每一个 Loader 都必须包含 test 和 use 两个选项。这段配置的意思是说，当 webpack 编译过程中遇到 require() 或 import 语句导入一个后缀名为 .css 的文件时，先将它通过 css-loader 转换，再通过 style-loader 转换，然后继续打包。use 选项的值可以是数组或字符串，如果是数组，它的编译顺序就是从后往前。

在 demo 目录下新建一个 style.css 的文件，并在 main.js 中导入：

```
/*style.css*/
#app{
    font-size: 50px;   /*定义字体大小*/
    color: blue;       /*定义蓝色字体*/
}
//main.js
import './style.css';
document.getElementById ('app') .innerHTML ='Hello webpack!';
```

重新执行 npm run dev 命令，可以看到页面中的文字已经变成蓝色，并且字号也大了，如图 9-4 所示。

图 9-4　页面效果

可以看到，CSS 是通过 JavaScript 动态创建 <style> 标签来写入的，这意味着样式代码都已经编译在 main.js 文件里，但在实际业务中，可能并不希望这样做，因为项目大了样式会很多，都放在 main.js 里太占体积，还不能做缓存。这时就要用到 webpack 最后一个重要的概念——插件（Plugins）。

webpack 的插件功能很强大而且可以定制。这里使用一个 mini-css-extract-plugin 的插件来把散落在各地的 CSS 提取出来，并生成一个 main.css 文件，最终在 index.html 里通过 <link> 的形式加载它。

通过 NPM 安装 mini-css-extract-plugin 插件：

```
npm install mini-css-extract-plugin -D
```

然后在配置文件中导入插件，并改写 loader 的配置：

```
//导入插件
let MiniCssExtractPlugin = require("mini-css-extract-plugin");
var config={
   //…
    //css文件处理
    module:{
        rules:[
            {
                test: /\.css$/,
                use : [
                    MiniCssExtractPlugin.loader,
                    { loader: "css-loader" }
                ]
            }
        ]
    },
    plugins: [
        //重命名提取后的 css 文件
        new MiniCssExtractPlugin({
            filename: "main.css"
        }),
    ]
};
```

插件还可以进行丰富的配置，本书会在后面结合 Vue 使用时详细介绍。现在重新启动服务，就可以看到 <style> 已经没有了，通过 <link> 引入的 main.css 文件已经生效。

webpack 虽然概念比较新，看似复杂，但它只不过是一个 JavaScript 配置文件，只要搞清楚入口（Entry）、出口（Output）、加载器（Loaders）和插件（Plugins）这 4 个概念，使用起来就不那么困难了。

9.3 单文件组件与 vue-loader

Vue 是一个渐进式的 JavaScript 框架，在使用 webpack 构建 Vue 项目时，可以使用一种新的构建模式：.vue 单文件组件。

顾名思义，.vue 单文件组件就是一个后缀名为 .vue 的文件，在 webpack 中使用 vue-loader 就可以对 .vue 格式的文件进行处理。

一个 .vue 文件一般包含 3 部分，即 <template>、<script> 和 <style>，如下所示：

```
<template></template>
<script>
    export default {
```

```
        name: "demo"
    }
</script>
<style scoped></style>
```

在component.vue文件中，<template></template>之间的代码就是该组件的模板HTML，<style></style>之间的是CSS样式，示例中的<style>标签使用了scoped属性，表示当前的CSS只在这个组件有效，如果不加，那么div的样式会应用到整个项目。<style>还可以结合CSS预编译一起使用，例如使用Less处理可以写成<style lang='less'>。

使用.vue文件需要先安装vue-loader、vue-style-loader等加载器并做配置。因为要使用ES 6语法，还需要安装babel和babel-loader等加载器。使用NPM逐个安装以下依赖：

```
npm install --save vue
npm install --save-dev vue-loader
npm install --save-dev vue-style-loader
npm install --save-dev vue-template-compiler
npm install --save-dev vue-hot-reload-api
npm install --save-dev babel
npm install --save-dev babel-loader
npm install --save-dev babel-core
npm install --save-dev babel-plugin-transform-runtime
npm install --save-dev babel-preset-env
npm install --save-dev babel-runtime
```

安装完成后，修改配置文件webpack.config.js来支持对.vue文件及ES 6的解析。需要使用插件VueLoaderPlugin在webpack.config.js里用const VueLoaderPlugin= require('vue-loader/lib/plugin')引入；然后在module.exports对象里添加plugins:[new VueLoaderPlugin()]。具体如下：

```
var path=require('path');
let MiniCssExtractPlugin = require("mini-css-extract-plugin");
const VueLoaderPlugin = require('vue-loader/lib/plugin')
var config={
    entry:{
        main:'./main'
    },
    output:{
        path:path.join(__dirname,'./dist'),
        publicPath:'/dist/',
        filename:'main.js'
    },
    module:{
        rules:[
            {
                test: /\.vue$/,
                loader: 'vue-loader'
            },
            {
                test:/\.js$/,
                loader:'babel-loader',
                exclude:/node_modules/
```

```
                },
                {
                    test: /\.css$/,
                    use : [
                        MiniCssExtractPlugin.loader,
                        { loader: "css-loader" }
                    ]
                }
            ]
        },
        plugins: [
            //重命名提取后的 css 文件
            new MiniCssExtractPlugin({
                filename: "main.css"
            }),
            new VueLoaderPlugin()
        ]
};
module.exports=config;
```

vue-loader 在编译 .vue 文件时，会对 <template>、<script>、<style> 分别处理，所以在 vue-loader 选项里多了一项 options 来进一步对不同语言进行配置。例如在对 CSS 进行处理时，会先通过 css-loader 解析，然后把处理结果再交给 vue-style-loader 处理。当用户的技术栈多样化时，可以给 <template>、<script> 和 <style> 指定不同的语言，例如 <template lang='jade'> 和 <style lang='less'>，然后配置 loaders 就可以了。

在 demo 目录下新建一个名为 .babelrc 的文件，并写入 babel 的配置，webpack 会依赖此配置文件来使用 babel 编译 ES 6 代码：

```
{
    "presets":["env"],
    "plugins":["transform-runtime"],
    "comments":false
}
```

配置好这些后，就可以使用 .vue 文件了。每个 .vue 文件就代表一个组件，组件之间可以相互依赖。

在 demo 目录下新建一个 app.vue 的文件并写入以下内容：

```
<template>
    <div> Hello {{frame}}</div>
</template>
<script>
    export default {
        data(){
            return {
                frame:'Vue.js'
            }
        }
    }
</script>
<style scoped>
    div{
        color:blue;
        font-size: 50px;
    }
</style>
```

提示

data (){} 等同于 data: function () {}。

在 <template> 内写的 HTML 语句的写法与 html 文件完全一样，不用加“\”换行，webpack 最终会把它编译为 Render 函数的形式。写在 <style> 里的样式，我们已经用插件 extract-text-webpack-plugin 配置过了，最终会统一提取并打包在 main.css 里，因为加了 scoped 属性，这部分样式只会对当前组件 app.vue 有效。

.vue 的组件是没有名称的，在父组件使用时可以对它自定义。写好了组件，就可以在入口 main.js 中使用它了。打开 main.js 文件，把内容替换为下面的代码：

```
//导入Vue框架
import Vue from 'vue';
//导入app.vue组件
import App from './app.vue';
//创建Vue实例
new Vue({
    el:"#app",
    render:h=>h(App)
})
```

提示

=> 是箭头函数 ,render: h => h(App) 等同于：

render: function (h) { return h(App);

也等同于：

render : h => { return h (App);

箭头函数里的 this 指向与普通函数是不一样的，箭头函数体内的 this 对象就是定义时所在的对象，而不是使用时所在的对象，例如：

```
function Timer(){
    this.id=1;
    var _this=this;
    setTimeout(function () {
        console.log(this.id); //undefined
        console.log(_this.id); //1
    },1000);
    setTimeout(()=>{
        console.log(this.id);  //1
    },1000);
}
var timer=new Timer();
```

执行命令 npm run dev，第一个 Vue 工程就跑起来了。打开 Chrome（谷歌浏览器）调试工具，在 Elements 面板可以看到，<div id="app"></div> 已经被组件替换，如图 9-5 所示。

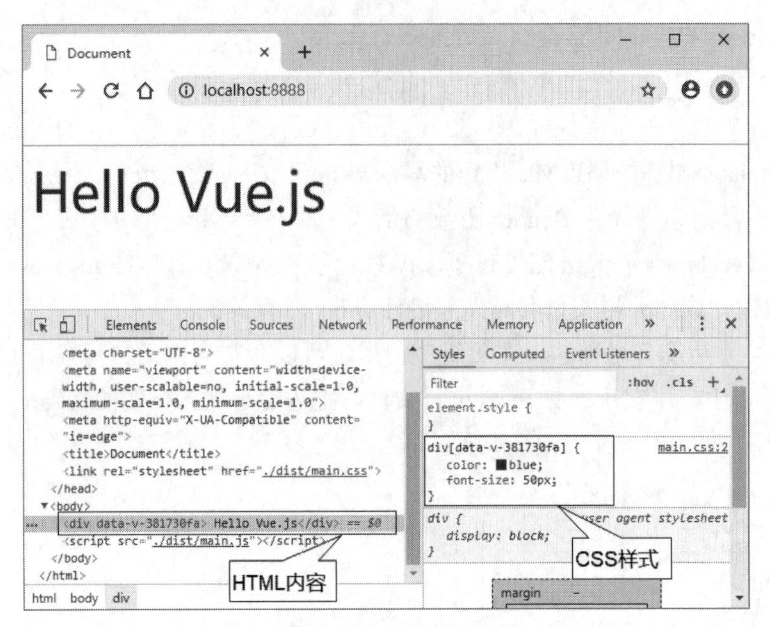

图 9-5　替换效果

之所以多了一串 data-v-xxx 内容，是因为使用了＜ style scoped ＞功能，如果去掉 scoped，就只剩下＜ div>Hello Vue.js</div ＞了。

接下来，在 demo 目录下再新建两个组件：title.vue 和 button.vue。

title.vue 组件：

```
<template>
    <h4>{{title}}</h4>
</template>
<script>
    export default{
        props:{
            title:{
                type:String
            }
        }
    }
</script>
<style scoped>
    h4{
        color:red;
    }
</style>
```

button.vue 组件：

```
<template>
    <button @click="handleClick" :style="styles">
        <slot></slot>
    </button>
</template>
<script>
    export default{
```

```
        props:{
            color:{
                type:String,
                default:'green'
            }
        },
        computed:{
            styles(){
                return{
                    background:this.color
                }
            }
        },
        methods:{
            handleClick(e){
                this.$emit('click',e);
            }
        }
    }
</script>
<style scoped>
    button{
        border: 0;
        color:white;
        padding:8px 12px;
    }
    button:active{
        position:relative;
        top:10px;
        left:10px;
    }
</style>
```

改写根实例 app.vue 组件，把 title.vue 和 button.vue 导入进来：

```
<template>
    <div>
        <v-title title="标题"></v-title>
        <v-button @click="handleClick()">单击按钮</v-button>
    </div>
</template>
<script>
    //导入组件
    import vTitle from  './title.vue';
    import vButton from  './button.vue';
    export default {
        components:{
            vTitle,
            vButton
        },
        methods:{
            handleClick(){
                alert('单击按钮了');
            }
        }
    }
</script>
```

提示

ES 6 语法提示。对象字面量缩写，当对象的 key 和 value 名称一致时，可以缩写成一个。例如：

```
components:{                    components:{
    vTitle,        等价于:          vTitle:vTitle,
    vButton                         vButton:vButton
},                              }
```

打开谷歌浏览器，已经正确渲染出了这两个组件，如图 9-6 所示。

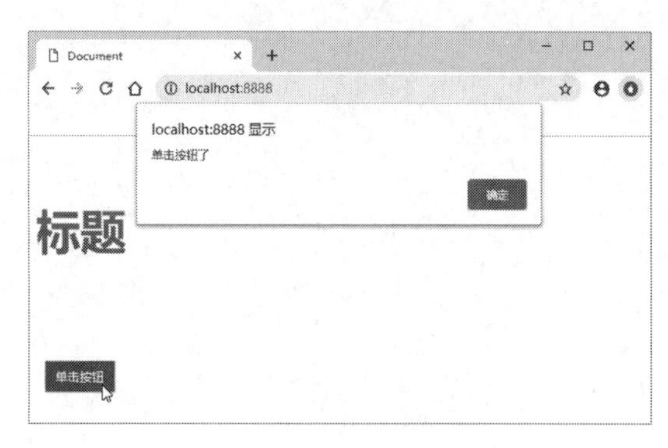

图 9-6　组件渲染

9.4 疑难解惑

疑问 1：在 webpack 4 中使用 extract-text-webpack-plugin 插件提取单独打包 CSS 文件时报错，提示这个插件要依赖 webpack 3 的版本，怎么解决？

在 webpack 4 中得使用 mini-css-extract-plugin 这个插件来单独打包 CSS 文件，具体的请看本章关于 mini-css-extract-plugin 的使用说明。

疑问 2：使用 webpack 4 来构建 Vue 2 的项目环境遇到了一些内置插件失效和环境依赖库的更新的地方，webpack 4 中环境及依赖库的版本要求是多少？

环境及依赖库的部分版本要求如表 9-1 所示。

表 9-1　部分环境及依赖库的版本

loader	最低版本要求	功能说明
vue-loader	15.0.0	解析、编译 Vue 单文件组件
vue-style-loader	4.1.0	解析、编译 Vue 单文件组件中的样式
babel-loader	7.1.3	对最新的 ES 语法进行转换
file-loader	1.1.10	批量修改文件路径，或者指定编译后文件存储路径
eslint-loader	2.0.0	代码检查

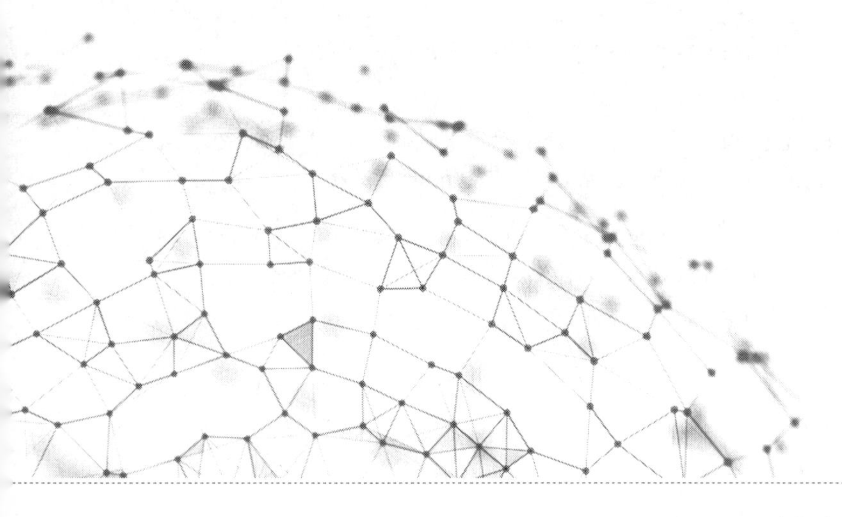

第10章

项目脚手架vue-cli

vue-cli 是一个官方发布的 Vue 项目脚手架，使用 vue-cli 可以快速创建 Vue 项目。本章将会从实际开发的角度，介绍项目脚手架整个搭建的过程和应用技巧。

10.1 脚手架的组件

脚手架致力于将 Vue 生态中的工具基础标准化。它确保了各种构建工具能够基于智能的默认配置平稳衔接，这样，开发人员可以专注在撰写应用上，而不必花好几天去纠结配置的问题。与此同时，它也为每个工具提供了调整配置的灵活性，无须 eject。

vue-cli 有几个独立的部分——如果了解过 Vue 的源代码，会发现这个仓库里同时管理了多个单独发布的包。

1. CLI

CLI（@vue/cli）是一个全局安装的 NPM 包，提供了终端里的 Vue 命令。它可以通过 vue create 命令快速创建一个新项目的脚手架，或者直接通过 vue serve 命令构建新想法的原型。也可以使用 vue ui 命令，通过一套图形化界面管理所有项目。

2. CLI 服务

CLI 服务（@vue/cli-service）是一个开发环境依赖。它是一个 NPM 包，局部安装在每个 @vue/cli 创建的项目中。

CLI 服务是构建于 webpack 和 webpack-dev-server 之上的，它包含以下内容。

(1) 加载其他 CLI 插件的核心服务。

(2) 一个针对绝大部分应用优化过的内部的 webpack 配置。

(3) 项目内部的 vue-cli-service 命令，提供 serve、build 和 inspect 命令。

@vue/cli-service 实际上大致等价于 react-scripts，尽管功能集合不一样。

3.CLI 插件

CLI 插件是向 Vue 项目提供可选功能的 NPM 包，例如 Babel/TypeScript 转译、ESLint 集成、单元测试和 end-to-end 测试等。Vue-cli 插件的名字以 @vue/cli-plugin-（内置插件）或 vue-cli-plugin-（社区插件）开头，使用非常方便。在项目内部运行 vue-cli-service 命令时，它会自动解析并加载 package.json 中列出的所有 CLI 插件。

插件可以作为项目创建过程的一部分，或在后期加入到项目中。它们也可以被归成一组可复用的 preset。

10.2 脚手架环境搭建

新版本的脚手架包名称由 vue-cli 改成了 @vue/cli。如果已经全局安装了旧版本的 Vue-cli (1.x 或 2.x)，需要先通过 npm uninstall vue-cli -g 或 yarn global remove vue-cli 卸载它。Vue-cli 需要 Node.js 8.9 或更高版本。

首先在 IE 浏览器中打开 node.js 官网 https://nodejs.org /en/，下载推荐版本。下载完成文件如图 10-1 所示，双击进行安装。

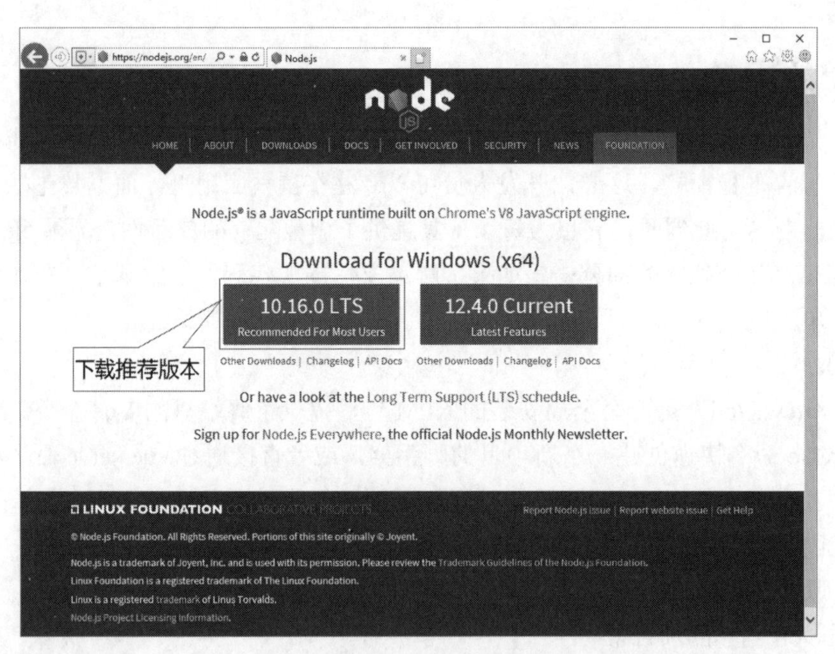

图 10-1　node 官网

安装过程很简单，一直单击 Next 按钮就可以了。具体的流程如图 10-2 所示。

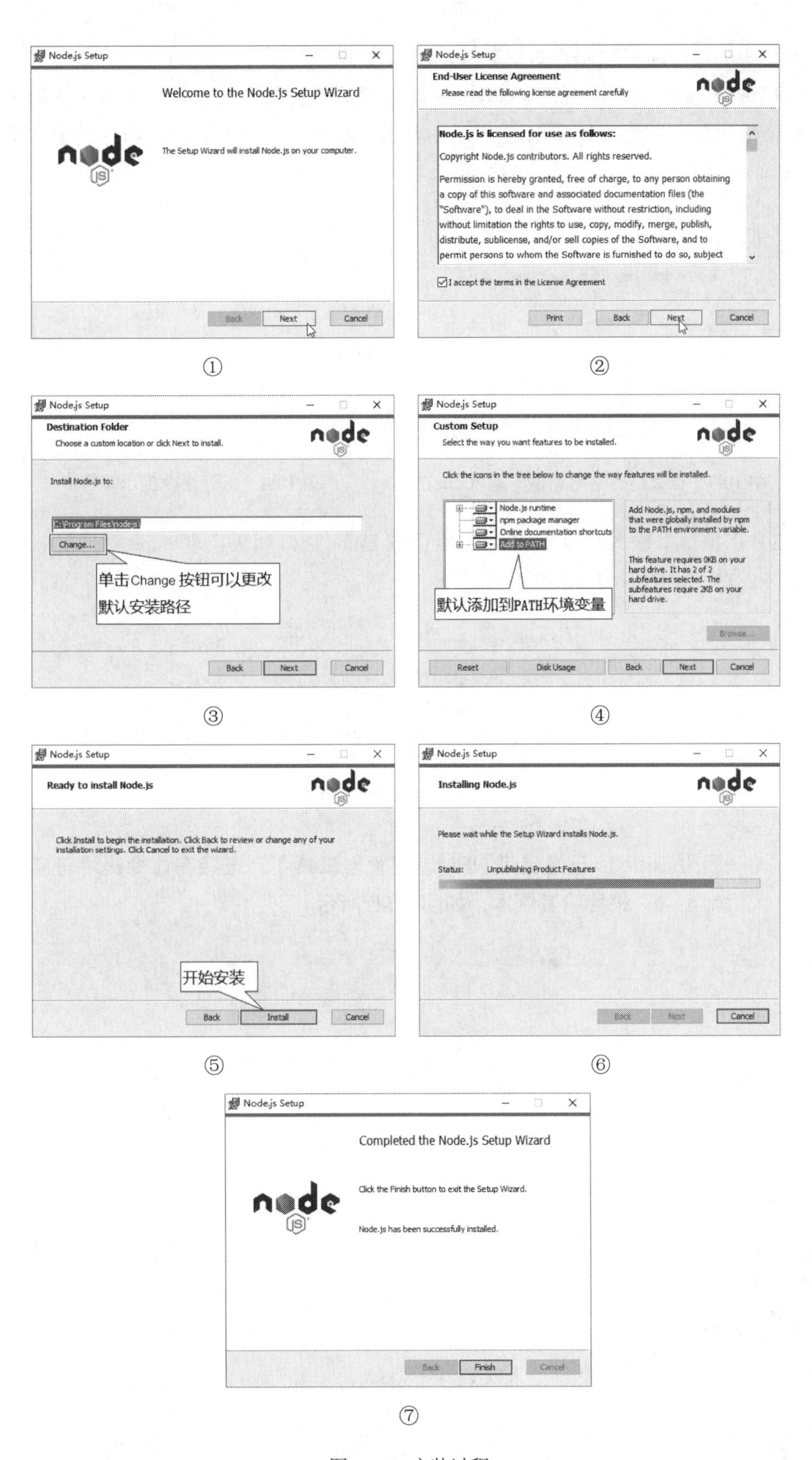
Node.js Setup
Welcome to the Node.js Setup Wizard
The Setup Wizard will install Node.js on your computer.
Back
Next
Cancel
①
End-User License Agreement
Please read the following license agreement carefully
Node.js is licensed for use as follows:
Copyright Node.js contributors. All rights reserved.
Permission is hereby granted, free of charge, to any person obtaining a copy of this software and associated documentation files (the "Software"), to deal in the Software without restriction, including without limitation the rights to use, copy, modify, merge, publish, distribute, sublicense, and/or sell copies of the Software, and to permit persons to whom the Software is furnished to do so, subject
I accept the terms in the License Agreement
Print
②
Destination Folder
Choose a custom location or click Next to install.
Install Node.js to:
C:\Program Files\nodejs\
Change...
单击Change按钮可以更改默认安装路径
③
Custom Setup
Select the way you want features to be installed.
Click the icons in the tree below to change the way features will be installed.
Node.js runtime
npm package manager
Online documentation shortcuts
Add to PATH
Add Node.js, npm, and modules that were globally installed by npm to the PATH environment variable.
This feature requires 0KB on your hard drive. It has 2 of 2 subfeatures selected. The subfeatures require 2KB on your hard drive.
默认添加到PATH环境变量
Browse...
Reset
Disk Usage
④
Ready to install Node.js
Click Install to begin the installation. Click Back to review or change any of your installation settings. Click Cancel to exit the wizard.
开始安装
Install
⑤
Installing Node.js
Please wait while the Setup Wizard installs Node.js.
Status: Unpublishing Product Features
⑥
Completed the Node.js Setup Wizard
Click the Finish button to exit the Setup Wizard.
Node.js has been successfully installed.
Finish
⑦

图 10-2　安装过程

安装成功后，需要检测是否安装成功。具体步骤如下。

（1）打开命令提示符窗口。按 window+R 组合键打开“运行”对话框，然后在“运行”对话框中输入 cmd，如图 10-3 所示；单击“确定”按钮即可打开命令提示符窗口，如图 10-4 所示。

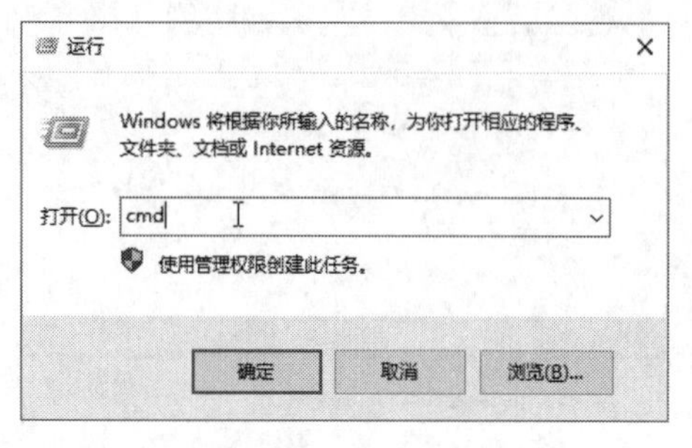

图 10-3　在“运行”对话框中输入 cmd

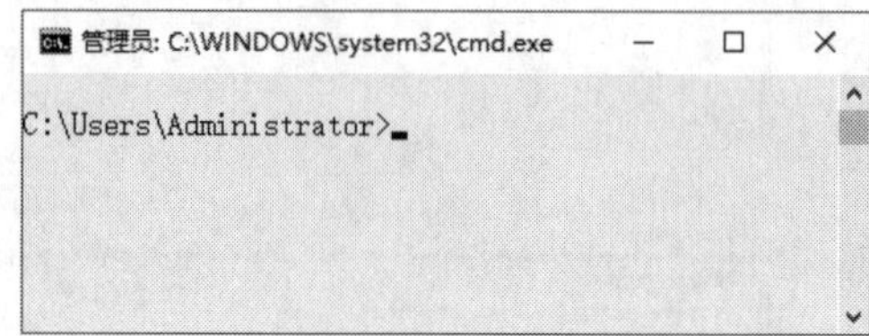

图 10-4　打开命令提示符窗口

（2）在窗口中输入命令 node -v，然后按 Enter 键，如果出现 node 对应的版本号，说明安装成功，如图 10-5 所示。

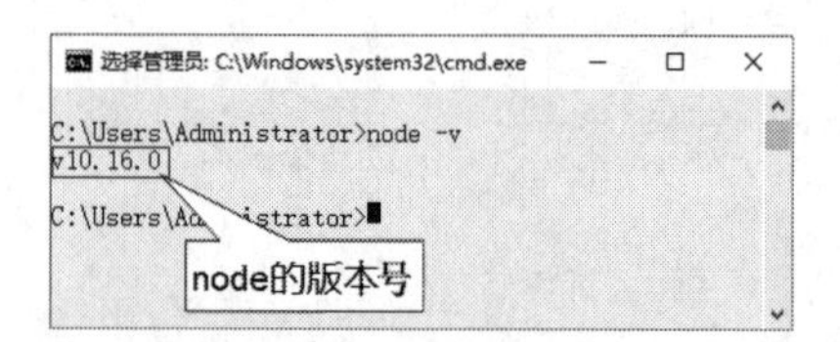

图 10-5　检查 node 版本

提示

因为 node.js 已经自带 NPM（包管理工具），直接在命令提示符窗口中输入“npm -v”来检验其版本，如图 10-6 所示。

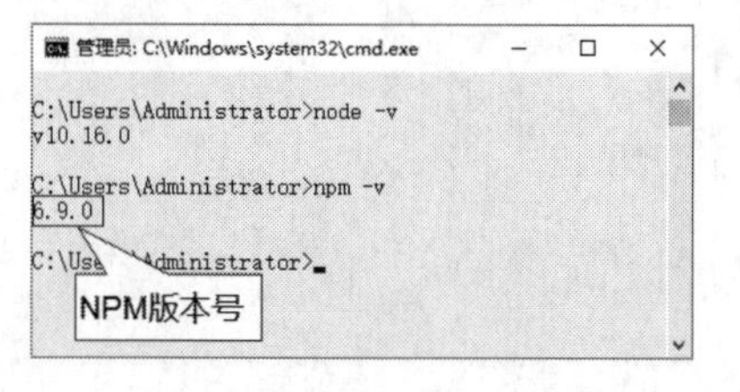

图 10-6　检查 npm 版本

10.3 安装脚手架

可以使用下列其中一个命令来安装脚手架：

```
npm install -g @vue/cli
```

或者

```
yarn global add @vue/cli
```

这里我们使用 npm install -g @vue/cli 命令来安装。在窗口中输入命令，按 Enter 键即可进行安装，如图 10-7 所示。

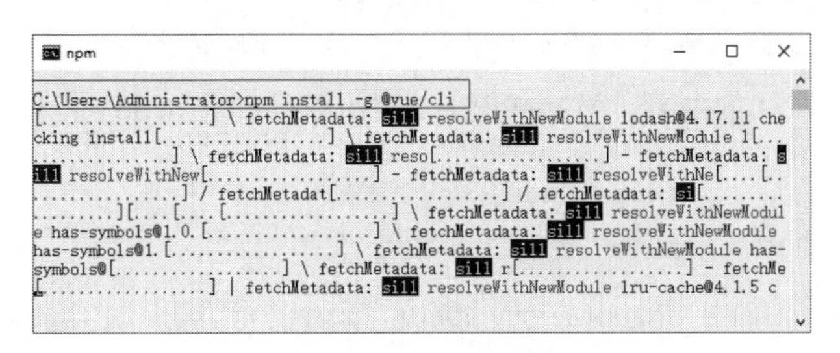

图 10-7　安装脚手架

安装之后，可以使用 vue --version 命令来检查其版本是否正确（3.x），如图 10-8 所示。

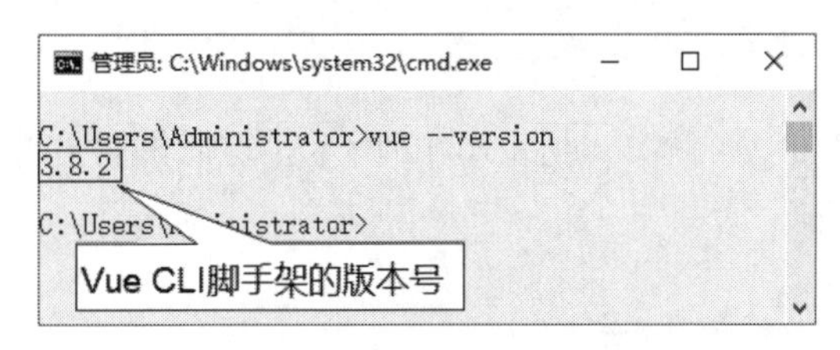

图 10-8　检查脚手架版本

10.4 创建项目

脚手架的环境配置完成后便可以使用脚手架来快速创建项目。

10.4.1　使用命令

首先要打开创建项目的路径，例如在桌面创建项目（C:\Users\Administrator\Desktop\），项目名称为 Hello，具体步骤如下。

（1）打开命令提示符窗口，在窗口中输入 cd C:\Users\Administrator\Desktop\ 命令，按 Enter 键进入到桌面目录，如图 10-9 所示。

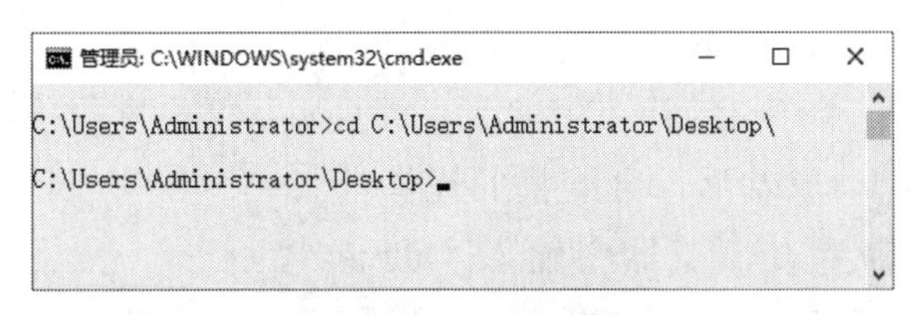

图 10-9　打开项目路径

注意

项目的名称不能大写，否则无法创建。

（2）在桌面创建hello项目。在窗口中输入vue create hello命令，按Enter键进行创建。紧接着系统会提示我们，是选择默认的配置，还是手动配置，如图10-10所示。

这里我们选择默认的配置，直接按Enter键，即可创建hello项目，如图10-11所示。

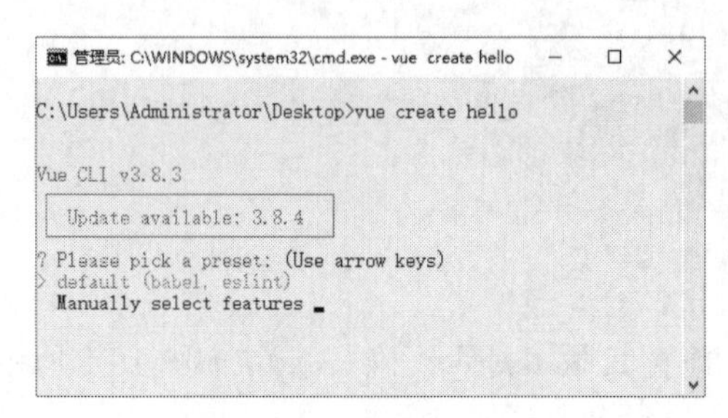

图10-10　选择配置方式

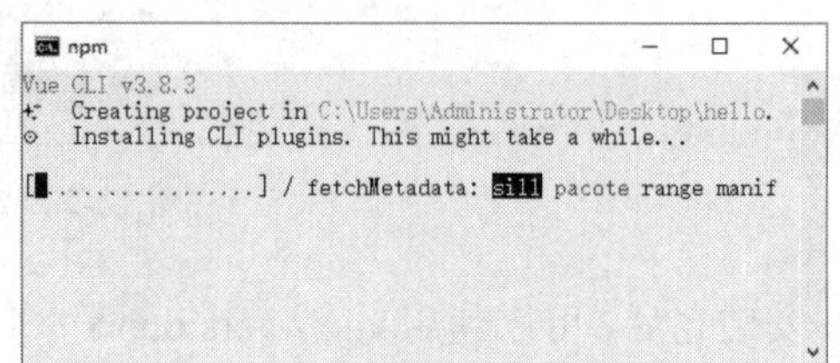

图10-11　创建hello项目

项目创建完成后，命令行窗口如图10-12所示。这时即可在桌面上看见创建的项目文件夹，如图10-13所示。

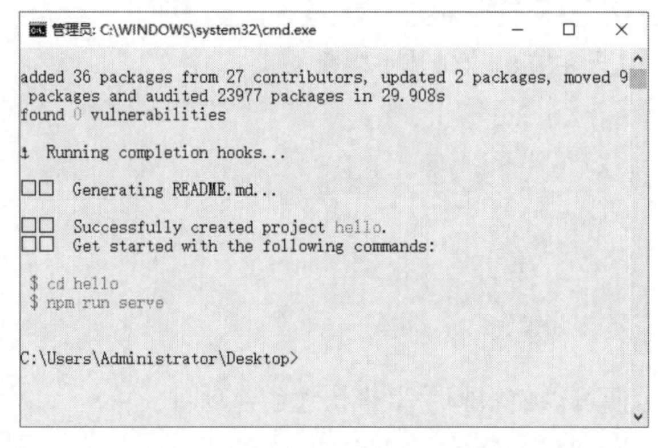

图10-12　项目创建完成

图10-13　项目文件夹

双击打开桌面上的hello文件夹，目录结构如图10-14所示。在项目中，可以根据习惯对该目录进行改造，在本书的后面案例中，就进行了改造。

（3）项目创建完成后，可以启动项目。紧接着上面的步骤，使用cd hello命令进入到项目，然后使用脚手架提供的npm run serve命令启动项目，如图10-15所示。

可以看到项目启动成功后，会给我们提供本地的测试域名，只需要在浏览器中输入“http://localhost:8080/”，即可打开项目，如图10-16所示。

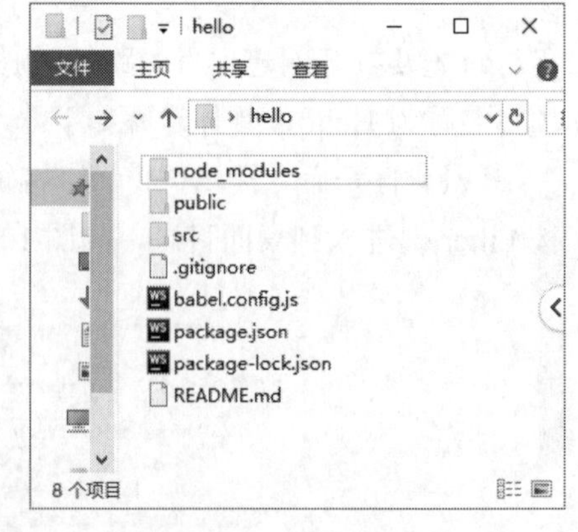

图10-14　项目目录结构

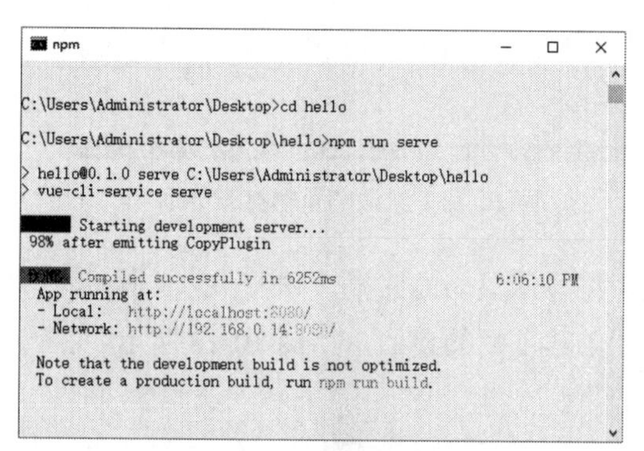

图 10-15　启动项目

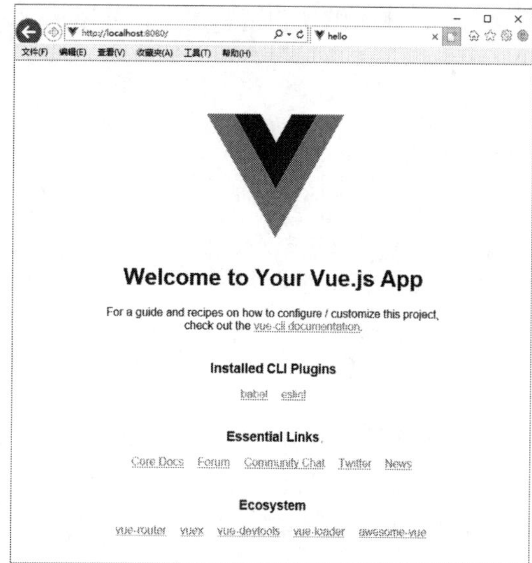

图 10-16　在浏览器中打开项目

vue create 命令有一些可选项，可以通过运行以下命令进行探索：

```
vue create --help
```

选项说明如下。

- -p, --preset <presetName>：忽略提示符并使用已保存的或远程的预设选项。
- -d, --default：忽略提示符并使用默认预设选项。
- -i, --inlinePreset <json>：忽略提示符并使用内联的 JSON 字符串预设选项。
- -m, --packageManager <command>：在安装依赖时使用指定的 NPM 客户端。
- -r, --registry <url>：在安装依赖时使用指定的 npm registry。
- -g, --git [message]：强制 / 跳过 git 初始化，并可选地指定初始化提交信息。
- -n, --no-git：跳过 git 初始化。
- -f, --force：覆写目标目录可能存在的配置。
- -c, --clone：使用 git clone 获取远程预设选项。
- -x, --proxy：使用指定的代理创建项目。
- -b, --bare：创建项目时省略默认组件中的新手指导信息。
- -h, --help：输出使用帮助信息。

10.4.2　使用图形化界面

可以通过 vue ui 命令，以图形化界面创建和管理项目。这里我们还是在桌面上进行创建，项目名称为“hello1”。

具体步骤如下。

（1）打开命令提示符窗口，在窗口中输入 cd C:\Users\Administrator\Desktop\ 命令，按 Enter 键进入到桌面。然后在窗口中输入 vue ui 命令，按 Enter 键，如图 10-17 所示，紧接着会在本地默认的浏览器上打开图形化界面，如图 10-18 所示。

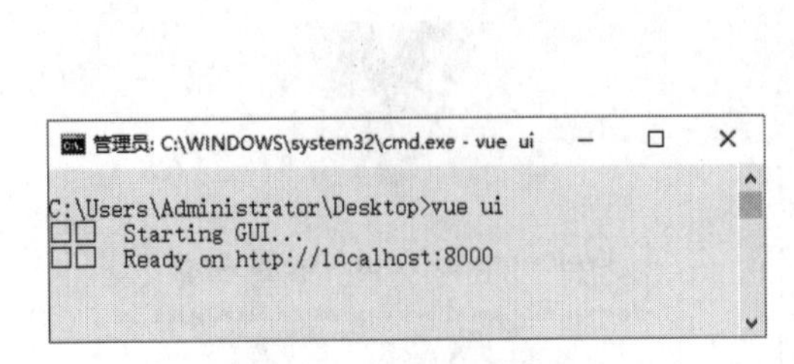

图 10-17　启动图形化界面

图 10-18　默认浏览器打开界面

（2）在图形化界面中单击“创建”按钮，将显示桌面的路径，如图 10-19 所示。

（3）单击“在此创建新项目”按钮，显示创建项目的界面，如图 10-20 所示。在详情选项中，根据需要进行选择。

图 10-19　单击“创建”按钮

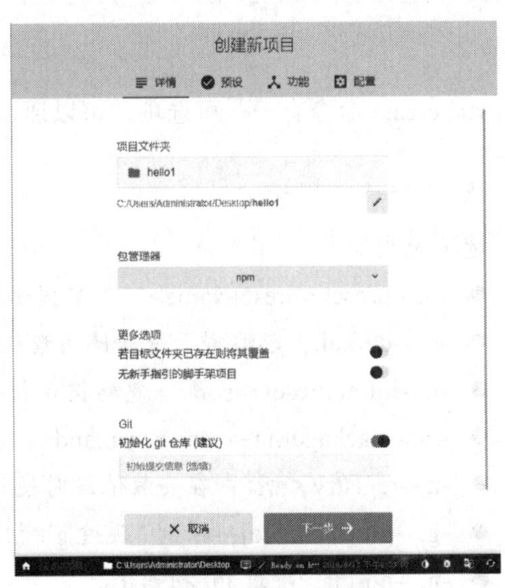

图 10-20　详情选项配置

（4）选择完成后，单击“下一步”按钮，将显示“预设”选项，如图 10-21 所示。根据需要选择一套预设即可。

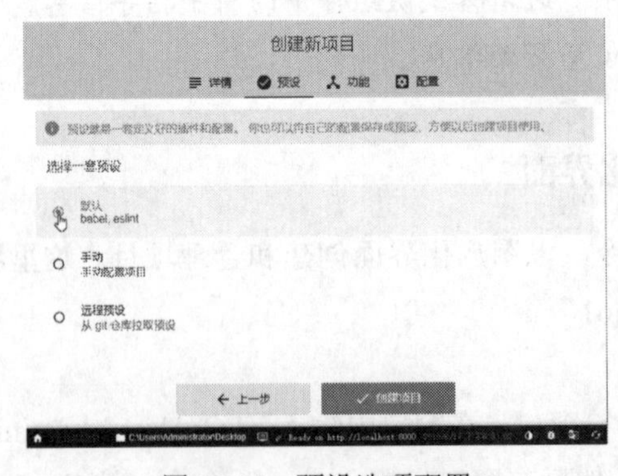

图 10-21　预设选项配置

（5）单击“创建项目”按钮，即可开始创建项目，如图 10-22 所示。

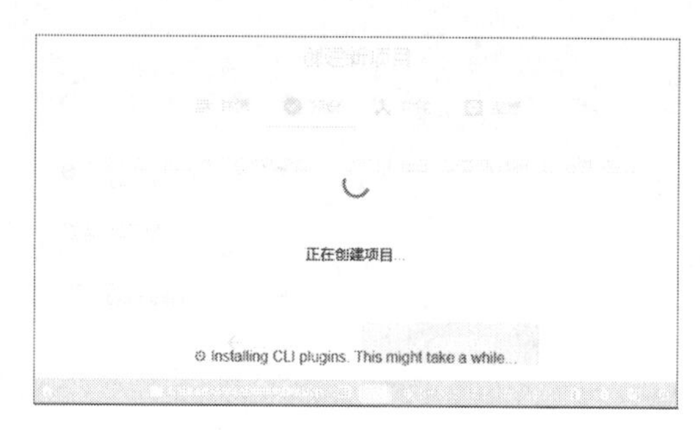

图 10-22　开始创建项目

（6）创建完成后，桌面上即可看到 hello1 项目的文件夹。浏览器中将显示如图 10-23 所示的界面，其他四个部分：插件、依赖、配置和任务，分别如图 10-24 ～图 10-27 所示。

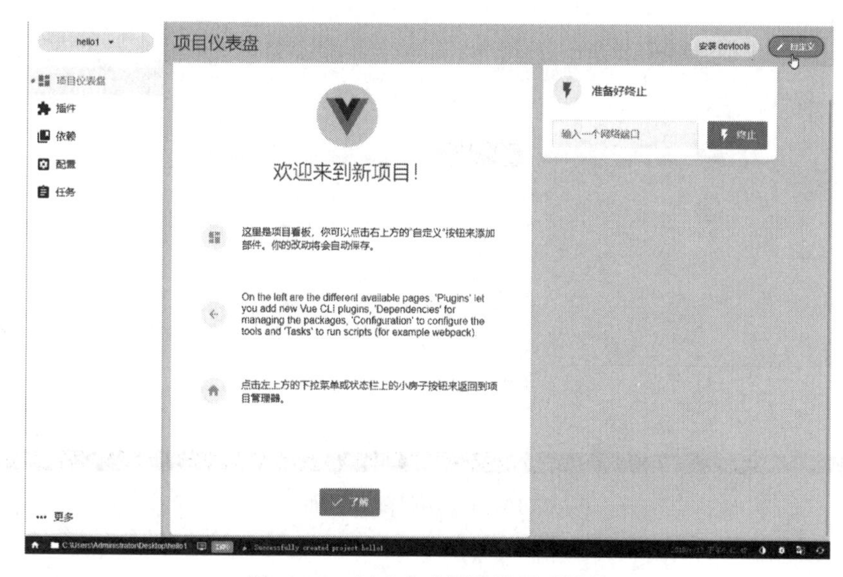

图 10-23　创建完成浏览器显示效果

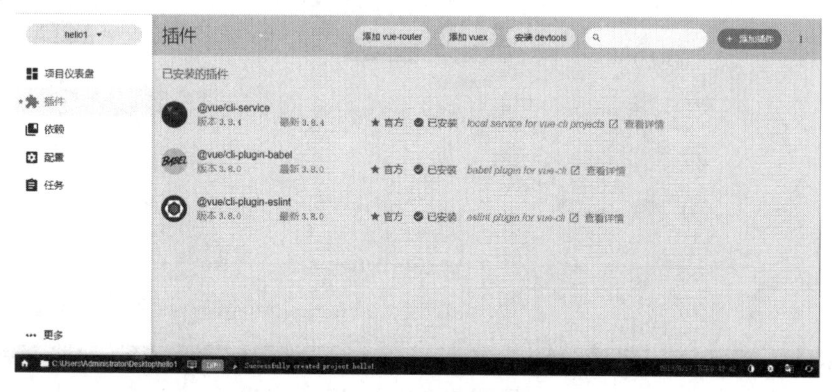

图 10-24　插件配置界面

图 10-25　依赖配置界面

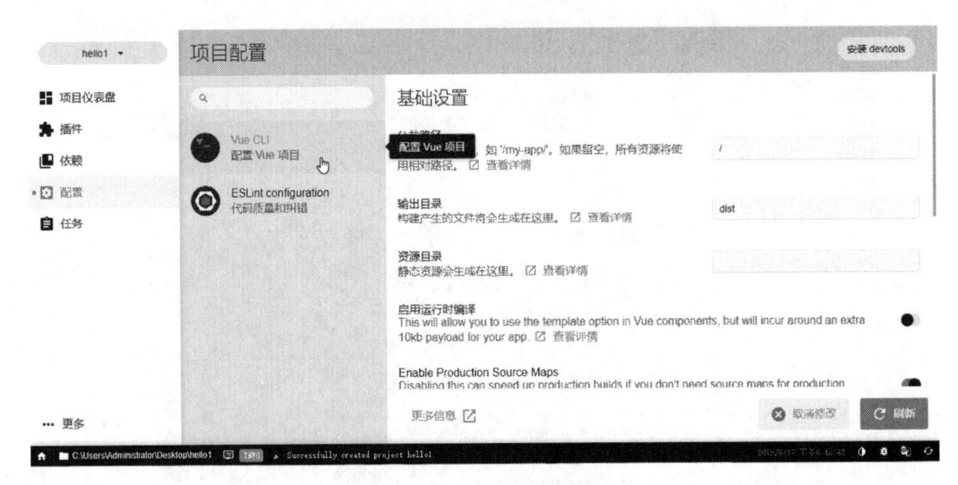

图 10-26　项目配置界面

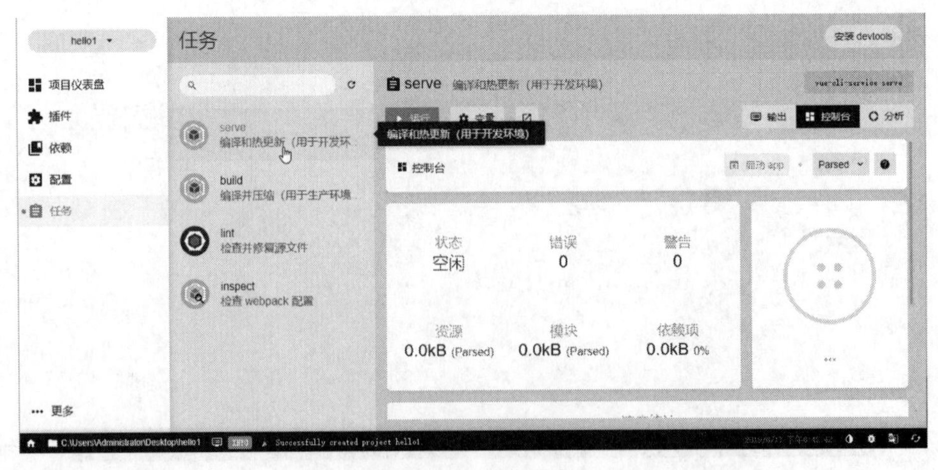

图 10-27　任务界面

10.5 疑难解惑

疑问 1：静态资源处理：assets 和 static 文件夹的区别是什么？

Vue-cli 有两个放置静态资源的地方，分别是 src/assets 文件夹和 static 文件夹。assets 目录中的文件会被 webpack 处理解析为模块依赖，只支持相对路径形式。例如：在 <img src="./logo.png" 和 background: url(./logo.png) 中，"./logo.png" 是相对的资源路径，将由 Webpack 解析为模块依赖。

static 目录下的文件并不会被 Webpack 处理，它们会直接被复制到最终的打包目录（默认是 dist/static）下。必须使用绝对路径引用这些文件，这是通过在 config.js 文件中的 build.assetsPublicPath 和 build.assetsSubDirectory 连接来确定的。

任何放在 static 文件中的数据，需要以绝对路径的形式引用：/static/[filename]。

在实际的开发中，static 用于放不会变动的文件，assets 用于放可能会变动的文件。

疑问 2：使用 Vue-cli 3.x 创建的项目如何实现跨域请求？

Vue-cli 创建的项目，可以直接利用 Node.js 代理服务器，通过修改 vue proxyTable 接口实现跨域请求。在 Vue-cli 项目中的 config 文件夹下的 index.js 配置文件中，修改前的 dev：

```
dev: {
    env: require('./dev.env'),
    port: 8080,
    autoOpenBrowser: true,
    assetsSubDirectory: 'static',
    assetsPublicPath: '/',
    proxyTable: {},
    cssSourceMap: false
  }
```

只要在 proxyTable:{} 项中配置参数，便可实现跨域。配置如下：

```
proxyTable: {
      '/api': {
          target: 'http://baidu.com',  //跨域的地址
          changeOrigin: true,          //是否跨域
          secure: false,               //如果是https接口，需要配置这个参数
          pathRewrite: {
               '^/api': '/api'//路径的替换规则，这里的配置是正则表达式，以/api开头的将会
//被用'/api'替换掉，假如后台文档的接口是/api/list/xxx，前端api接口写：axios.get('/api/
//list//xxx')，被处理之后实际访问的是：http:// baidu.com/api/list/xxx
          },
      }
}
```

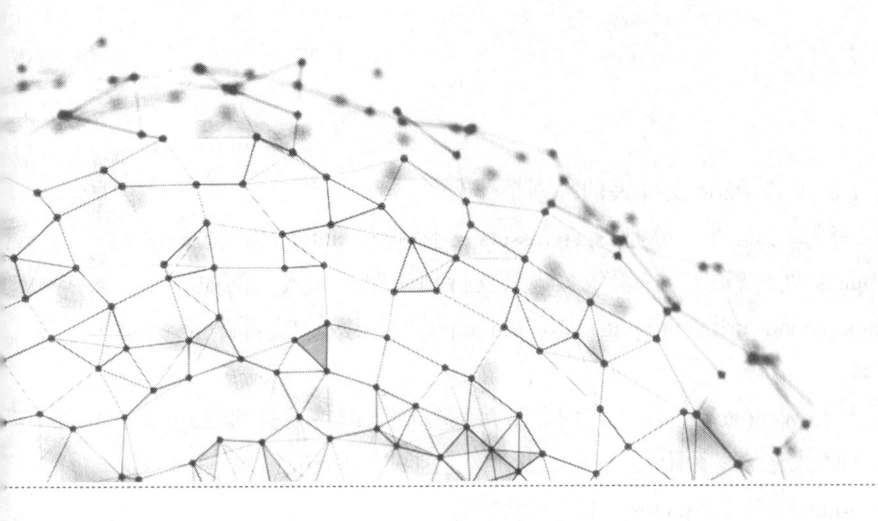

第11章 前端路由技术

前端路由是什么？如果之前从事的是后端的工作，或者虽然有接触前端，但是并没有使用到单页面应用的话，对于这个概念来说还是会很陌生的。为什么会在单页面应用中存在这么一个概念？前端路由与后端的路由有什么不同？本章将介绍前端路由的知识。

11.1 实现 Vue 前端路由控制

在传统的多页面应用中，网站的每一个 URL 地址都是对应于服务器磁盘上的一个实际物理文件。例如，当访问 https://www.yousite.com/index.html 这个网址的时候，服务器会自动把我们的请求对应到当前站点路径下面的 index.html 文件，然后再给予响应，将这个文件返回给浏览器。当跳转到别的页面上时，则会再重复一遍这个过程。

但是在单页面应用中，整个项目中只会存在一个 html 文件，当用户切换页面时，只是通过对这个唯一的 HTML 文件进行动态重写，从而达到响应用户的请求。也就是说，从切换页面这个角度上说，应用只是在第一次打开时请求了服务器。

因为访问的页面并不是真实存在的，所以如何正确地在一个 html 文件中展现出用户想要访问的信息，就成为单页面应用需要考虑的问题，而前端路由技术就很好地解决了这个问题。

11.1.1 前端路由的实现方式

前端路由的实现方式主要是通过 hash 路由匹配或者是采用 HTML 5 中的 history API，不管是 hash 路由还是使用 history 路由模式，其实都是基于浏览器自身的特性。

hash 路由在某些情况下，需要定位页面上的某些位置，就像下面的例子中展现的那样，通过锚点进行定位，单击不同的按钮跳转到不同的位置，这其实就是 hash。

```
<div class="container">
    <a class="btn" href="#image1">图片1</a>
    <a class="btn" href="#image2">图片2</a>
</div>
<img src="image1.jpg" id="image1">
<img src="image2.jpg" id="image2">
```

hash 路由是通过浏览器 location 对象中的 hash 属性实现的，它会记录链接地址中 '#' 后面的内容（:#part1）。因此，可以通过监听 window.onhashchange 事件获取到跳转前后访问的地址，从而实现地址切换的目的。

history 路由：在之前的 HTML 版本中，可以通过 history.back()、history.forward() 和 history.go() 方法来完成在用户历史记录中向后和向前的跳转。而 history 路由则是使用了 HTML 5 中新增的 pushState 事件和 replaceState() 事件。

通过这两个新增的 API，就可以实现无刷新地更改地址栏链接，配合 AJAX 就可以做到整个页面的无刷新跳转。

在 Vue 中，Vue Router 是官方提供的路由管理器，它和 Vue 的核心深度集成，不管是采用 hash 的方式还是使用 history API，对实现前端路由都有很好的支持，所以这里采用 Vue Router 这一组件来实现前端路由。

11.1.2 路由实现步骤

首先需要将 Vue Router 添加引用到项目中，这里采用直接引用 .js 文件的方式为我们的示例代码添加前端路由支持。

```
<script src="vue-router.js"></script>
```

在 Vue 中使用 Vue Router 构建单页面应用，只需要将组件 (components) 映射到定义的路由 (routes) 规则中，然后告诉 Vue Router 在哪里渲染它们，并将这个路由配置挂载到 Vue 实例节点上即可。

在 Vue Router 中，使用 router-link 标签来渲染链接。当然，默认生成的是 a 标签，如果想要将路由信息生成别的 HTML 标签，则可以使用 tag 属性指明需要生成的标签类型。

```
<!-- 默认渲染成a标签 -->
router-link to="/home">首页</router-link>
<!--渲染成 button标签-->
< <router-link to="/list" tag="button">列表</router-link>
```

可以看到，当我们指定 tag 属性为 button 后，页面渲染后的标签就变成了 button 按钮，如图 11-1 所示。

图 11-1　页面渲染后的标签

从图 11-1 中可以看出，当前的链接地址为 #/home，也就是说，通过 router-link 生成的标签，当页面地址与对应的路由规则匹配成功后，将自动设置 class 属性值为 .router-link-active。也可以通过指定 active-class 属性或者在构造 VueRouter 对象时使用 linkActiveClass 来自定义链接激活时使用的 CSS 类名。

```
<!-- 使用属性来设定自定义激活类名 -->
<router-link to="/home" active-class="style">首页</router-link>
<!-- 在构造对象时设定全局默认类名 -->
const router = new VueRouter({
    routes: [],
    linkActiveClass: style
})
```

当路由表构建完成后，对于指向路由表中的链接，需要在页面上找一个地方去显示已经渲染完成后的组件，这时就需要使用 router-view 标签去告诉程序，需要将渲染后的组件显示在当前位置。

在下面的示例代码中，模拟了 Vue 中路由的使用过程，当访问 #/home 时会加载 home 组件，而当链接跳转到 #/list 时，则会加载 list 组件。在 list 组件中又包含了两个子路由，通过单击 list 组件中的子路由地址，从而加载对应的 login 组件和 register 组件。

在上面的代码中，也使用到了嵌套路由和路由的重定向。通过使用路由重定向，用户访问网站时可以定向到 /home 路由，而嵌套路由则可以将 URL 中各段动态路径也按某种结构对应到实际嵌套的各层组件上。

例如，这里的 login 组件和 register 组件，它们都是位于 list 组件中的。因此，在构建 url 时，应该将该地址位于 /list url 后面，从而更好地表达这种关系。所以，在 list 组件中添加了一个 router-view 标签，用来渲染出嵌套的组件内容。同时，通过定义 routes 时，在参数中使用 children 属性，从而达到配置嵌套路由信息的目的。

【例 11.1】路由实现示例。

```
<style>
        #app{
            text-align: center;
        }
        .container {
            background-color: #73ffd6;
            margin-top: 20px;
            height: 300px;
        }
        .son{
            margin-top: 30px;
        }
</style>
<body>
<div id="app">
    <!-- 通过 router-link 标签来生成导航链接 -->
    <router-link to="/home">首页</router-link>
    <router-link to="/list">列表</router-link>
    <div class="container">
        <!-- 将选中的路由渲染到 router-view 下-->
        <router-view></router-view>
    </div>
</div>
<template id="tmpl">
    <div>
        <h3>列表内容</h3>
        <!-- 生成嵌套子路由地址 -->
        <router-link to="/list/login">登录</router-link>
        <router-link to="/list/register">注册</router-link>
        <div class="son">
            <!-- 生成嵌套子路由渲染节点 -->
            <router-view></router-view>
        </div>
    </div>
</template>
<script>
    // 1.定义路由跳转的组件模板
    const home = {
        template: '<div><h3>首页内容</h3></div>'
    }
    const list = {
        template: '#tmpl'
    }
    const login = {
        template: '<div> 登录页面内容</div>'
    }

    const register = {
        template: '<div>注册页面内容</div>'
    }
    // 2.定义路由信息
    const routes = [
        //路由重定向：当路径为/时，重定向到/home路径
        {
            path: '/',
            redirect: '/home'
        },
```

```
            {
                path: '/home',
                component: home
            },
            {
                path: '/list',
                component: list,
                //嵌套路由
                children: [
                    {
                        path: 'login',
                        component: login
                    },
                    {
                        path: 'register',
                        component: register
                    },
                    //当路径为/list时，重定向到/list/login路径
                    {
                        path: '/list',
                        redirect: '/list/login'
                    }
                ]
            }
        ]
        const router = new VueRouter({
            //mode: 'history', //使用 history 模式还是hash路由模式
            routes
        })
        // 3.挂载到当前 Vue 实例上
        const app = new Vue({
            el: '#app',
            data:{},
            methods: {},
            router: router
        });
</script>
</body>
```

在 IE 11 浏览器里面运行，单击“列表”链接，然后单击“注册”链接，效果如图 11-2 所示。

图 11-2　路由实现

11.2 命名路由、命名视图和路由传参

在上一节中，简单介绍了前端路由的知识以及如何在 Vue 中使用 Vue Router 来实现前端路由。但是在实际使用中，经常会遇到路由传参或者一个页面是由多个组件组成的情况。本节就来介绍在这两种情况下，Vue Router 的使用方法以及一些可能涉及的知识。

11.2.1 命名路由

在某些时候，生成的路由 URL 地址可能会很长，在使用中可能会显得有些不便。这时候通过一个名称来标识一个路由会更方便一些。因此在 Vue Router 中，可以在创建 Router 实例的时候，通过在 routes 配置中给某个路由设置名称，从而方便地调用路由。

```
const router = new VueRouter({
  routes:[
    {
      path: '/form',
      name: 'form',
      component: '<div>form组件</div>'
    }
  ]
})
```

命名路由之后，当需要使用 router-link 标签进行跳转时，就可以采取给 router-link 的 to 属性传一个对象的方式，跳转到指定的路由地址上，例如：

```
<router-link :to="{ name:'form'}">名称</router-link>
```

【例 11.2】命名路由示例。

```
 <style>
        #app{
            text-align: center;
        }
        .container {
            background-color: #73ffd6;
            margin-top: 20px;
            height: 300px;
        }
        .son{
            margin-top: 30px;
        }
    </style>
<body>
<div id="app">
    <!-- 通过 router-link 标签来生成导航链接 -->
    <router-link :to="{name:'router1'}">首页</router-link>
    <router-link :to="{name:'router2'}">列表</router-link>
    <div class="container">
        <!-- 将选中的路由渲染到 router-view 下-->
        <router-view></router-view>
```

```
        </div>
    </div>
    <template id="tmpl">
        <div>
            <h3>列表内容</h3>
            <router-link :to="{name:'router3'}">登录</router-link>
            <router-link :to="{name:'router4'}">注册</router-link>
            <div class="son">
                <!-- 生成嵌套子路由渲染节点 -->
                <router-view></router-view>
            </div>
        </div>
    </template>
    <script>
        // 1.定义路由跳转的组件模板
        const home = {
            template: '<div><h3>首页内容</h3></div>'
        }
        const list = {
            template: '#tmpl'
        }
        const login = {
            template: '<div> 登录页面内容</div>'
        }

        const register = {
            template: '<div>注册页面内容</div>'
        }
        // 2.定义路由信息
        const router = new VueRouter({
            //mode: 'history', //使用 history 模式还是 hash 路由模式
            routes:[
                {
                    path: '/',
                    redirect: '/home'
                },
                {
                    path: '/home',
                    name: 'router1',
                    component: home
                },
                {
                    path: '/list',
                    name: 'router2',
                    component: list,
                    //命名二级路由
                    children:[
                        {
                            path:'/login',
                            name:'router3',
                            component:login
                        },
                        {
                            path:'/list/register',
                            name:'router4',
                            component:register
                        },
                    ]
                }
```

```
        ]
    })
    // 3.挂载到当前 Vue 实例上
    const app = new Vue({
        el: '#app',
        data: {},
        methods: {},
        router: router
    });
</script>
</body>
```

在 IE 11 浏览器里面运行，效果与示例 11.1 相同。

也可以使用 params 属性设置参数，例如：

```
<router-link :to="{ name: 'user', params: { userId: 123 }}">User</router-link>
```

还可以用代码调用 router.push()：

```
router.push({ name: 'user', params: { userId: 123 }})
```

这两种方式都会把路由导航到 /user/123 路径。

11.2.2 命名视图

整个页面可能是由多个 Vue 组件所构成的，例如，后台管理首页可能是由 sidebar(侧导航)、header（顶部导航）和 main(主内容）这三个 Vue 组件构成的。此时，通过 Vue Router 构建路由信息时，如果一个 URL 只能对应一个 Vue 组件，整个页面肯定是无法正确显示的。

在上一节的学习中，在构建路由信息的时候，使用到两个特殊的标签：router-view 和 router-link。通过 router-view 标签，可以指定组件渲染显示到什么位置。当需要在一个页面上显示多个组件的时候，就需要在页面中添加多个 router-view 标签。

那么，是不是可以通过一个路由对应多个组件，然后按需渲染在不同的 router-view 标签上呢？按照上一节关于 Vue Router 的使用方法，可以很容易实现下面的代码。

【例 11.3】 测试一个路由对应多个组件。

```
<body>
<div id="app">
    <router-view></router-view>
    <div class="container">
        <router-view></router-view>
        <router-view></router-view>
    </div>
</div>
<template id="sidebar">
    <div class="sidebar">
        侧边栏内容
    </div>
```

```
</template>
<script>
    // 1.定义路由跳转的组件模板
    const header = {
        template: '<div class="header"> 头部内容 </div>'
    }
    const sidebar = {
        template: '#sidebar',
    }
    const main = {
        template: '<div class="main">主要内容</div>'
    }
    // 2.定义路由信息
    const routes = [{
        path: '/',
        component: header
    }, {
        path: '/',
        component: sidebar
    }, {
        path: '/',
        component: main
    }]
    const router = new VueRouter({
        routes
    })
    // 3.挂载到当前 Vue 实例上
    const vm = new Vue({
        el: '#app',
        data: {},
        methods: {},
        router: router
    });
</script>
</body>
```

在 IE 11 浏览器里面运行，效果如图 11-3 所示。

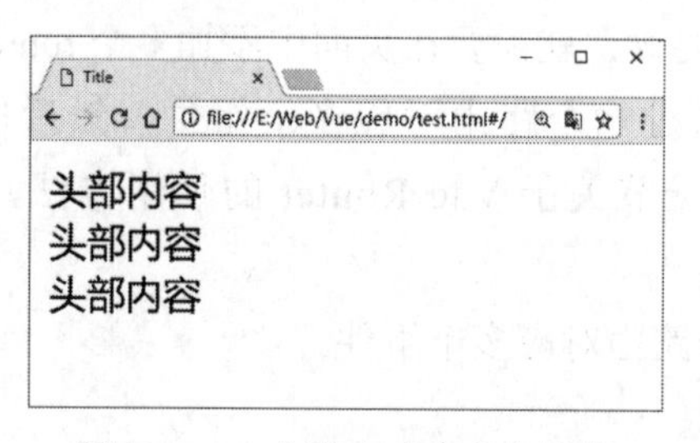

图 11-3　一个路由对应多个组件

可以看到，并没有实现我们想要实现的效果，将一个路由信息对应到多个组件时，不管有多少个 router-view 标签，程序都会将第一个组件渲染到所有的 router-view 标签上。

在 Vue Router 中，可以通过命名视图的方式实现一个路由信息按照需要去渲染到页面中指定的 router-view 标签。

命名视图与命名路由的实现方式相似，命名视图通过在 router-view 标签上设定 name 属性，之后在构建路由与组件的对应关系时，以一种 name:component 的形式构造出一个组件对象，从而指明是在哪个 router-view 标签上加载什么组件。

注意

在指定路由对应的组件时，使用的是 components（包含 s）属性进行配置组件。

实现命名视图的代码如下。

```
<div id="app">
    <router-view></router-view>
    <div class="container">
        <router-view name="sidebar"></router-view>
        <router-view name="main"></router-view>
    </div>
</div>
<script>
    //定义路由信息
    const routes = [{
        path: '/',
        components: {
            default: header,
            sidebar: sidebar,
            main: main
        }
    }]
</script>
```

在 router-view 中，默认的 name 属性值为 default，所以这里的 header 组件对应的 router-view 标签就可以不设定 name 属性值。

【例 11.4】命名视图示例。

```
<style>
        .style1{
            height: 20vh;
            background: #0BB20C;
            color: white;
        }
        .style2{
            background: #9e8158;
            float: left;
            width: 30%;
            height: 70vh;
            color: white;
        }
        .style3{
            background: #2d309e;
            float: left;
            width: 70%;
            height: 70vh;
            color: white;
        }
    </style>
<body>
<div id="app">
    <div class="style1">
```

```
            <router-view></router-view>
        </div>
        <div class="container">
            <div class="style2">
                <router-view name="sidebar"></router-view>
            </div>
            <div class="style3">
                <router-view name="main"></router-view>
            </div>
        </div>
    </div>
    <template id="sidebar">
        <div class="sidebar">
            侧边栏导航内容
        </div>
    </template>
    <script>
        // 1.定义路由跳转的组件模板
        const header = {
            template: '<div class="header"> 头部内容 </div>'
        }
        const sidebar = {
            template: '#sidebar'
        }
        const main = {
            template: '<div class="main">正文部分</div>'
        }
        // 2.定义路由信息
        const routes = [{
            path: '/',
            components: {
                default: header,
                sidebar: sidebar,
                main: main
            }
        }]
        const router = new VueRouter({
            routes
        })
        // 3.挂载到当前 Vue 实例上
        const vm = new Vue({
            el: '#app',
            data: {},
            methods: {},
            router: router
        });
    </script>
    </body>
```

在 IE 11 浏览器里面运行，效果如图 11-4 所示。

图 11-4　命名视图

11.2.3　路由传参

在很多情况下，例如表单提交、组件跳转之类的操作，需要使用到上一个表单、组件的一些数据，这时就需要将需要的参数通过参数传参的方式在路由间进行传递。下面介绍两种传参方式：query 传参和 param 传参。

1. query 传参

query 查询参数传参，就是将需要的参数以 key=value 的方式放在 url 地址中。

在下面的示例中，想要实现通过单击 main 组件中的子组件 form 组件上的按钮，将表单的内容传递到 info 子组件中进行显示。

例如这里的需求，需要 info 组件中显示上一个页面的数据，所以 info 页面显示的 URL 地址应该为 /info?email=xxx&password=xxx，这里的 email 和 password 参数值则是 form 组件上用户输入的值。之后通过获取这两个参数值即可实现我们的目标。

将实例化的 Vue Router 对象挂载到 Vue 实例后，Vue Router 在 Vue 实例上就创建了两个属性对象：this.$router（router 实例）和 this.$route（当前页面的路由信息）。可以通过 vm.$route 获取到当前页面的路由信息，如图 11-5 所示。

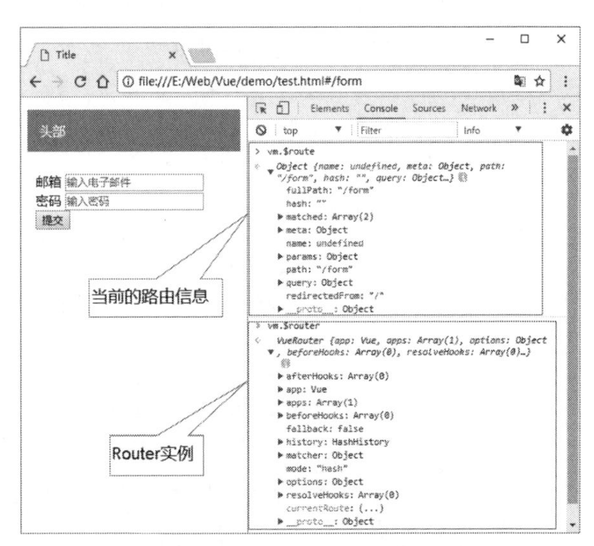

图 11-5　获取的当前页面的路由信息

这时就可以直接通过 $route.query 参数名的方式获取到对应的参数值。同时可以发现，fullPath 属性可以获取到当前页面的地址和 query 查询参数，而 path 属性则只是获取到当前的路由信息。

因为在使用 Vue Router 时已经将 Vue Router 实例挂载到 Vue 实例上，所以就可以直接通过调用 $router.push

方法来导航到另一个页面，所以这里 form 组件中的按钮事件，就可以使用这种方式完成路由地址的跳转。

【例 11.5】 query 传参。

```
    <style>
        .style1{
            background: #0BB20C;
            color: white;
            padding: 15px;
            margin: 15px 0;
        }
        .main{
            padding: 10px;
        }
    </style>
<body>
<div id="app">
    <div>
        <div class="style1">
            <router-view></router-view>
        </div>
    </div>
    <div class="main">
        <router-view name="main"></router-view>
    </div>
</div>
<template id="sidebar">
    <div>
        <ul>
             <router-link v-for="(item,index) in menu" :key="index" :to="item.
               url" tag="li">{{item.name}}
            </router-link>
        </ul>
    </div>
</template>

<template id="main">
    <div>
        <router-view></router-view>
    </div>
</template>
<template id="form">
    <div>
        <form>
            <div>
                <label for="exampleInputEmail1">邮箱</label>
                 <input type="email" id="exampleInputEmail1" placeholder="输入电
                   子邮件" v-model="email">
            </div>
            <div>
                <label for="exampleInputPassword1">密码</label>
                <input type="password" id="exampleInputPassword1" placeholder=
                   "输入密码" v-model="password">
            </div>
            <button type="submit" @click="submit">提交</button>
        </form>
    </div>
```

```
</template>
<template id="info">
    <div class="card" style="margin-top: 5px;">
        <div class="card-header">
            输入的信息
        </div>
        <div class="card-body">
            <blockquote class="blockquote mb-0">
                <p>邮箱：{{ $route.query.email }} </p>
                <p>密码：{{ $route.query.password }}</p>
            </blockquote>
        </div>
    </div>
</template>
<script>
    // 1.定义路由跳转的组件模板
    const header = {
        template: '<div class="header"> 头部 </div>'
    }
    const sidebar = {
        template: '#sidebar',
        data:function() {
            return {
                menu: [{
                    name: 'Form',
                    url: '/form'
                },
                 {
                    name: 'Info',
                    url: '/info'
                }]
            }
        },
    }
    const main = {
        template: '#main'
    }
    const form = {
        template: '#form',
        data:function() {
            return {
                email: '',
                password: ''
            }
        },
        methods: {
            submit() {
                this.$router.push({
                   path: '/info?email=' + this.email + '&password=' + this.password
                })
            }
        },
    }
    const info = {
        template: '#info'
    }
    // 2.定义路由信息
    const routes = [{
        path: '/',
```

```
            components: {
                default: header,
                sidebar: sidebar,
                main: main
            },
            children: [{
                path: '',
                redirect: 'form'
            },
            {
                path: 'form',
                component: form
            },
            {
                path: 'info',
                component: info
            }]
        }]
        const router = new VueRouter({
            routes
        })
        // 3.挂载到当前 Vue 实例上
        const vm = new Vue({
            el:'#app',
            data:{},
            methods:{},
            router:router
        });
    </script>
    </body>
```

在 IE 11 浏览器里面运行，在邮箱中输入“123456”，在密码中输入“abcdefg”，单后单击“提交”按钮，内容传递到 info 子组件中进行显示，效果如图 11-6 所示。

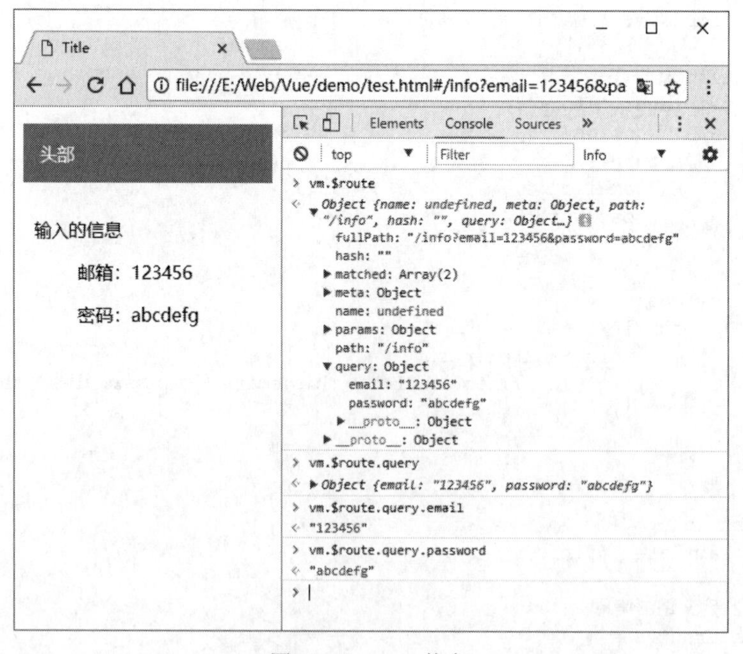

图 11-6　query 传参

2. param 传参

与获取 query 参数的方式相同，同样可以通过 vm.$route 获取到当前路由信息，从而在 param 对象中，通过 $route.params 参数名的方式，获取到通过 param 的方式进行参数传递的值。不过，与 query 查询参数传参不同的是，在定义路由信息时，需要以占位符（: 参数名）的方式将需要传递的参数指定到路由地址中，实现代码如下：

```
const routes=[{
        path:'/',
        components:{
            default: header,
            sidebar: sidebar,
            main: main
        },
        children: [{
            path: '',
            redirect: 'form'
        }, {
            path: 'form',
            name: 'form',
            component: form
        }, {
            path: 'info/:email/:password',
            name: 'info',
            component: info
        }]
    }]
```

因为在使用 $route.push 进行路由跳转时，如果提供了 path 属性，则对象中的 params 属性会被忽略，所以这里可以采用命名路由的方式进行跳转或者直接将参数值传递到路由 path 路径中。与使用 query 查询参数传递参数不同，这里的参数如果不进行赋值的话，就无法与匹配规则对应，也就无法跳转到指定的路由地址中。

```
const form = {
        template: '#form',
        data:function() {
            return {
                email: '',
                password: ''
            }
        },
        methods: {
            submit:function() {
                //方式1
                this.$router.push({
                    name: 'info',
                    params: {
                        email: this.email,
                        password: this.password
                    }
                })
                //方式2
                this.$router.push({
                    path: '/info/${this.email}/${this.password}',
                })
```

```
            }
        },
    }
```

其余部分的代码与使用 query 传参的代码相同，具体的实现示例如下。

【例 11.6】param 传参示例。

```
<style>
        .style1{
            background: #0BB20C;
            color: white;
            padding: 15px;
            margin: 15px 0;
        }
        .main{
            padding: 10px;
        }
    </style>
<body>
<div id="app">
    <div>
        <div class="style1">
            <router-view></router-view>
        </div>
    </div>
    <div class="main">
        <router-view name="main"></router-view>
    </div>
</div>
<template id="sidebar">
    <div>
        <ul>
            <router-link v-for="(item,index) in menu" :key="index" :to="item.
                url" tag="li">{{item.name}}
            </router-link>
        </ul>
    </div>
</template>

<template id="main">
    <div>
        <router-view></router-view>
    </div>
</template>
<template id="form">
    <div>
        <form>
            <div>
                <label for="exampleInputEmail1">邮箱</label>
                <input type="email" id="exampleInputEmail1" placeholder="输入电
                  子邮件" v-model="email">
            </div>
            <div>
                <label for="exampleInputPassword1">密码</label>
                <input type="password" id="exampleInputPassword1" placeholder=
                  "输入密码" v-model="password">
            </div>
            <button type="submit" @click="submit">提交</button>
```

```
        </form>
    </div>
</template>
<template id="info">
    <div>
        <div>
            输入的信息
        </div>
        <div>
            <blockquote>
                <p>邮箱：{{ $route.params.email }} </p>
                <p>密码：{{ $route.params.password }}</p>
            </blockquote>
        </div>
    </div>
</template>
<script>
    // 1.定义路由跳转的组件模板
    const header = {
        template: '<div class="header">头部</div>'
    }
    const sidebar = {
        template: '#sidebar',
        data:function() {
            return {
                menu: [{
                    displayName: 'Form',
                    routeName: 'form'
                }, {
                    displayName: 'Info',
                    routeName: 'info'
                }]
            }
        },
    }
    const main = {
        template: '#main'
    }
    const form = {
        template: '#form',
        data:function() {
            return {
                email: '',
                password: ''
            }
        },
        methods: {
            submit:function() {
                //方式1
                this.$router.push({
                    name: 'info',
                    params: {
                        email: this.email,
                        password: this.password
                    }
                })
            }
        },
    }
```

```
        const info = {
            template: '#info'
        }
        // 2.定义路由信息
        const routes = [{
            path: '/',
            components: {
                default: header,
                sidebar: sidebar,
                main: main
            },
            children: [{
                path: '',
                redirect: 'form'
            }, {
                path: 'form',
                name: 'form',
                component: form
            }, {
                path: 'info/:email/:password',
                name: 'info',
                component: info
            }]
        }]
        const router = new VueRouter({
            routes
        })
        // 3.挂载到当前 Vue 实例上
        const vm = new Vue({
            el: '#app',
            data: {},
            methods: {},
            router: router
        });
    </script>
    </body>
```

在 IE 11 浏览器里面运行，在邮箱中输入“123456”，在密码中输入“abcdefg”，单后单击“提交”按钮，内容传递到 info 子组件中进行显示，效果如图 11-7 所示。

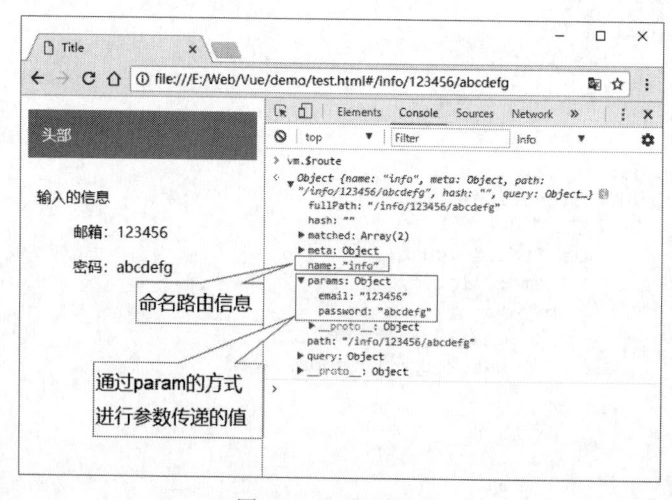

图 11-7 param 传参

11.3 编程式导航

在使用 Vue Router 时，通常会通过 router-link 标签去生成跳转到指定路由的链接，但是在实际的前端开发中，更多的是通过 JavaScript 的方式进行跳转。例如很常见的一个交互需求——用户提交表单，提交成功后跳转到上一页面，提交失败则留在当前页面。这时候如果还是通过 router-link 标签进行跳转就不合适了，需要通过 JavaScript 根据表单返回的状态进行动态判断。

在使用 Vue Router 时，已经将 Vue Router 的实例挂载到了 Vue 实例上，可以借助 $router 的实例方法，通过编写 JavaScript 代码的方式实现路由间的跳转，而这种方式就是一种编程式的路由导航。

在 Vue Router 中有三种导航方法，分别为 push、replace 和 go。最常见的通过在页面上设置 router-link 标签进行路由地址间的跳转，就等同于执行了一次 push 方法。

1. push 方法

当需要跳转新页面时，可以通过 push 方法将一条新的路由记录添加到浏览器的 history 栈中，通过 history 的自身特性，从而驱使浏览器进行页面的跳转。同时，因为在 history 会话历史中会一直保留着这个路由信息，所以后退时还是可以退回到当前的页面。

在 push 方法中，参数可以是一个字符串路径或者是一个描述地址的对象，它其实就等同于调用了 history.pushState 方法。

```
//字符串 => /first
this.$router.push('first')
//对象=> /first
this.$router.push({ path: 'first' })
//带查询参数=>/first?abc=123
this.$router.push({ path: 'first', query: { abc: '123' }})
```

注意

当传递的参数为一个对象并且 path 与 params 共同使用时，对象中的 params 属性不会起任何作用，需要采用命名路由的方式进行跳转，或者是直接使用带有参数的全路径。

```
const userId = '123'
// 使用命名路由 => /user/123
this.$router.push({ name: 'user', params: { userId }})
// 使用带有参数的全路径 => /user/123
this.$router.push({ path: '/user/${userId}' })
// 这里的 params 不生效 => /user
this.$router.push({ path: '/user', params: { userId }})
```

2. go 方法

当使用 go 方法时，可以在 history 记录中向前或者后退多少步，也就是说通过 go 方法可以在已经存储的 history 路由历史中来回跳。

```
//在浏览器记录中前进一步，等同于history.forward()
this.$router.go(1)
//后退一步记录，等同于history.back()
this.$router.go(-1)
//前进3步记录
this.$router.go(3)
```

3. replace 方法

replace 方法同样可以达到实现路由跳转的目的，从名字中可以看出，与使用 push 方法跳转不同是，使用 replace 方法时，并不会往 history 栈中新增一条新的记录，而是会替换掉当前的记录，因此无法通过后退按钮再回到被替换前的页面。

```
this.$router.replace({
    path: '/special'
})
```

下面通过编程式路由，实现路由间的切换，示例代码如下。

【例 11.7】实现路由间的切换。

```
<style>
        .style1{
            background: #0BB20C;
            color: white;
            height: 300px;
        }
    </style>
<body>
<div id="app">
    <div class="main">
        <div >
            <button @click="goFirst">第一页</button>
            <button @click="goSecond">第二页</button>
            <button @click="goThird">第三页</button>
            <button @click="goFourth">第四页</button>
            <button @click="pre">后退</button>
            <button @click="next">前进</button>
            <button @click="replace">替换当前页为特殊页</button>
        </div>
        <div class="style1">
            <router-view></router-view>
        </div>
    </div>
</div>

<script>
    const first = {
        template: '<h3>第一页的内容</h3>'
    };;
    const second = {
        template: '<h3>第二页的内容</h3>'
    };
    const third = {
        template: '<h3>第三页内容</h3>'
    };
```

```
const fourth = {
    template: '<h3>第四页的内容</h3>'
};
const special = {
    template: '<h3>特殊页面的内容</h3>'
};
const router = new VueRouter({
    routes: [
        {
            path: '/first',
            component: first
        },
        {
            path: '/second',
            component: second
        },
        {
            path: '/third',
            component: third
        },
        {
            path: '/fourth',
            component: fourth
        },
        {
            path: '/special',
            component: special
        }
    ]
});
const vm = new Vue({
    el: '#app',
    data: {},
    methods: {
        goFirst:function() {
            this.$router.push({
                path: '/first'
            })
        },
        goSecond:function() {
            this.$router.push({
                path: '/second'
            })
        },
        goThird:function() {
            this.$router.push({
                path: '/third'
            })
        },
        goFourth:function() {
            this.$router.push({
                path: '/fourth'
            })
        },
        next:function() {
            this.$router.go(1)
        },
        pre:function() {
            this.$router.go(-1)
```

```
            },
            replace:function() {
                this.$router.replace({
                    path: '/special'
                })
            }
        },
        router: router
    })
</script>
</body>
```

在 IE 11 浏览器里面运行，效果如图 11-8 所示。

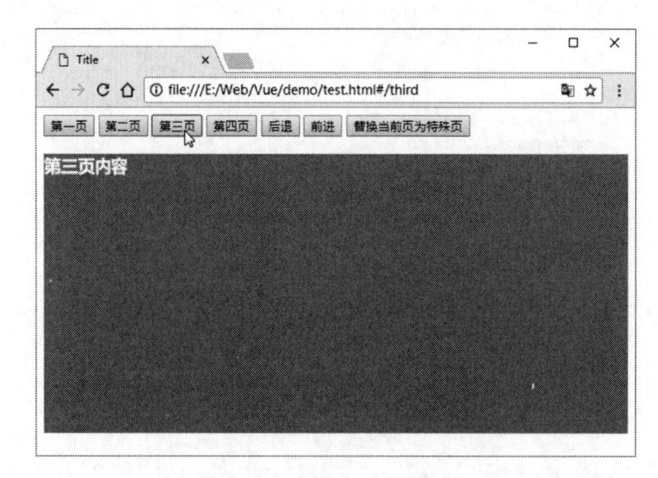

图 11-8　实现路由间的切换

11.4 组件与 Vue Router 间解耦

在使用路由传参的时候，将组件与 Vue Router 强绑定在了一起，这意味着在任何需要获取路由参数的地方，都需要加载 Vue Router，使组件只能在某些特定的 URL 上使用，限制了其灵活性。

在之前学习组件相关的知识时，提到了可以通过组件的 props 选项来实现子组件接收父组件传递的值。而在 Vue Router 中，同样提供了通过使用组件的 props 选项来进行解耦的功能。

11.4.1　布尔模式解耦

在下面的示例中，当定义路由模板时，指定需要传递的参数为 props 选项中的一个数据项，通过定义路由规则，指定 props 属性为 true，即可实现组件和 Vue Router 之间的解耦。

【例 11.8】布尔模式示例。

```
  <style>
        .style1{
            background: #0BB20C;
            color: white;
        }
    </style>
<body>
<div id="app">
    <div class="main">
        <div >
            <button @click="goFirst">第一页</button>
            <button @click="goSecond">第二页</button>
            <button @click="goThird">第三页</button>
            <button @click="goFourth">第四页</button>
            <button @click="pre">后退</button>
            <button @click="next">前进</button>
            <button @click="replace">替换当前页为特殊页</button>
        </div>
       <div class="style1">
           <router-view></router-view>
       </div>
    </div>
</div>
<script>
    const first = {
        template: '<h3>第一页的内容</h3>'
    };
    const second = {
        template: '<h3>第二页的内容</h3>'
    };
    const third = {
        props: ['id'],
        template: '<h3>第三页的内容---{{id}}</h3>'
    };
    const fourth = {
        template: '<h3>第四页的内容</h3>'
    };
    const special = {
        template: '<h3>特殊页面的内容</h3>'
    };
    const router = new VueRouter({
        routes: [
            {
                path: '/first',
                component: first
            },
            {
                path: '/second',
                component: second
            },
            {
                path: '/third/:id',
                component: third,
                props: true
            },
            {
```

```
                path: '/fourth',
                component: fourth
            },
            {
                path: '/special',
                component: special
            }
        ]
    });
    const vm = new Vue({
        el: '#app',
        data: {},
        methods: {
            goFirst:function() {
                this.$router.push({
                    path: '/first'
                })
            },
            goSecond:function() {
                this.$router.push({
                    path: '/second'
                })
            },
            goThird:function() {
                this.$router.push({
                    path: '/third'
                })
            },
            goFourth:function() {
                this.$router.push({
                    path: '/fourth'
                })
            },
            next:function() {
                this.$router.go(1)
            },
            pre:function() {
                this.$router.go(-1)
            },
            replace:function() {
                this.$router.replace({
                    path: '/special'
                })
            }
        },
        router: router
    })
</script>
</body>
```

在 IE 11 浏览器里面运行，选择第三页，在 url 路径中输入“/abcdefg”，然后按 Enter 键，效果如图 11-9 所示。

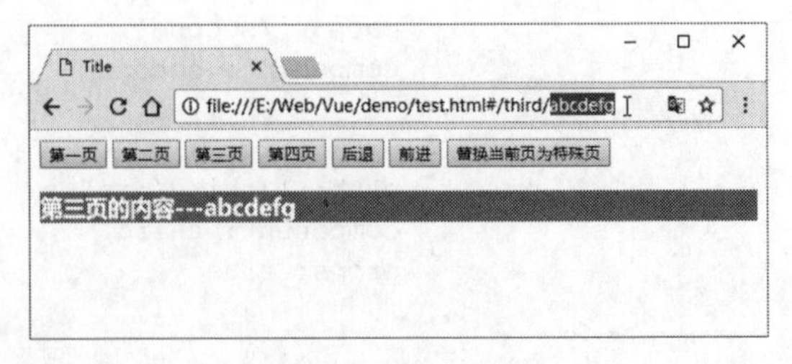

图 11-9　布尔模式

注意

上面示例采用 param 传参的方式进行参数传递，而在组件中并没有加载 Vue Router 实例，也完成了对于路由参数的获取。采用此方法，只能实现基于 param 方式进行传参的解耦。

11.4.2 对象模式解耦

针对定义路由规则时，指定 props 属性为 true 这一种情况，在 Vue Router 中，还可以给路由规则的 props 属性定义成一个对象或是函数。如果定义成对象或是函数，此时并不能实现对于组件以及 Vue Router 间的解耦。

将路由规则的 props 定义成对象后，此时不管路由参数中传递是任何值，最终获取到的都是对象中的值。需要注意的是，props 中的属性值必须是静态的，不能采用类似于子组件同步获取父组件传递的值作为 props 中的属性值。

下面的示例与上面示例中的代码类似，只需更改相应的代码即可。

【例 11.9】对象模式解耦。

```
<script>
    const third = {
        props: ['name'],
        template: '<h3>第三页的内容 ---{{name}} </h3>'
    }
    const router = new VueRouter({
        routes: [{
            path: '/third/:name',
            component: third,
            props: {
                name: 'xiaohong'
            }
        }]
    })
    const vm = new Vue({
        el: '#app',
        data: {},
        methods: {
            goThird() {
                this.$router.push({
                    path: '/third'
                })
            }
        },
        router: router
    })
</script>
```

在 IE 11 浏览器里面运行，选择第三页，在 url 路径中输入“/123456”，然后按 Enter 键，效果如图 11-10 所示。

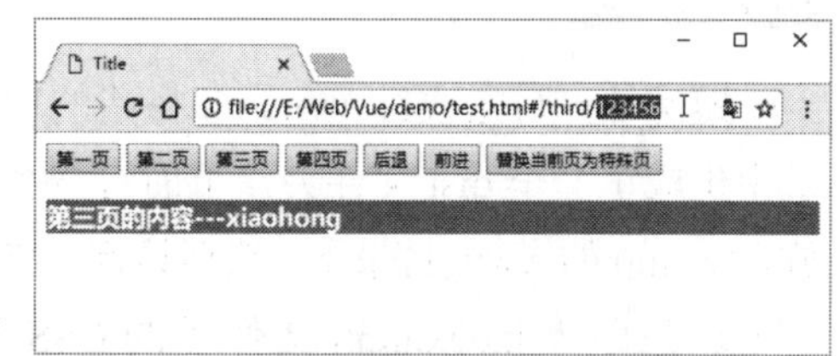

图 11-10　对象模式

11.4.3 函数模式解耦

在对象模式中，只能接收静态的props属性值，而当使用函数模式之后，就可以对静态值做数据的进一步加工或者是与路由传参的值进行结合。

下面的示例与上面示例中的代码类似，只需更改相应的代码即可。

【例 11.10】函数模式示例。

```
<script>
    const third = {
        props: ['name',"id"],
        template: '<h3>第三页的内容---{{name}}——{{id}}</h3>'
    };
    const router = new VueRouter({
        routes: [{
            path: '/third',
            component: third,
            props: (route)=>({
                id:route.query.id,
                name:"xiaohong"
            })
        }]
    })
    const vm = new Vue({
        el: '#app',
        data: {},
        methods: {
            goThird:function() {
                this.$router.push({
                    path: '/third'
                })
            },
        },
        router: router
    })
</script>
```

在IE 11浏览器里面运行，选择第三页，在url路径中输入“?id=123456”，然后按Enter键，效果如图11-11所示。

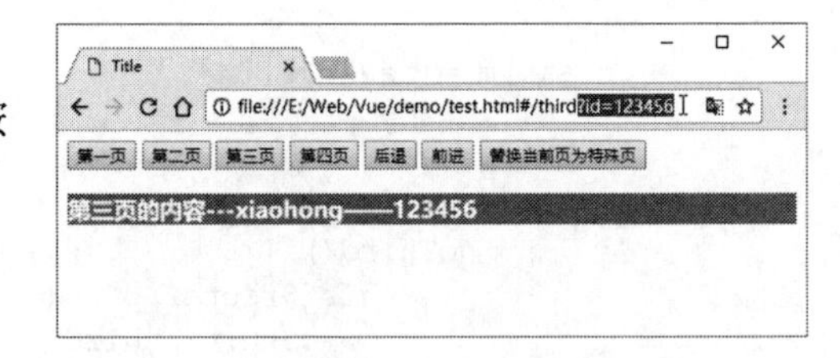

图 11-11　函数模式

11.5 疑难解惑

疑问 1：写完项目（vue-cli搭建的项目）直接将npm run build打包之后，在项目目录中生成了一个dist文件夹，里面有一个index.html和一个static文件夹，把dist文件夹放在wamp的www目录下，然后访问127.0.0.1/dist/，发现index.html内容为空，为什么？

这是因为index.html文件里的CSS和JS文件的路径不对导致的。打开项目中的config文件夹下的index.js文件：

```
build: {
    env: require('./prod.env'),
      //下面是相对路径的拼接，假如当前根目录是config，那么下面配置的index属性的属性值就是
     //dist/index.html
    index: path.resolve(__dirname, '../dist/index.html'),
    //下面定义的是静态资源的根目录，也就是dist目录
    assetsRoot: path.resolve(__dirname, '../dist'),
    //下面定义的是静态资源根目录的子目录static，也就是dist目录下面的static
    assetsSubDirectory: 'static',
    //下面定义的是静态资源的公开路径，也就是真正的引用路径
    assetsPublicPath: '/',
    …

}
```

只需要把 assetsPublicPath: '/' 中的“/”换成“./”即可，然后再执行 npm run build 打包命令，打包完成后可发现 index.html 文件能够访问了。

疑问 2：在项目中，使用 vue-router 时 URL 格式引发 404 的问题。

vue-router 提供一个 mode 参数，用来控制 URL 的格式，默认的是用 hash 格式，如果在项目中使用的是 history 格式：

```
const router = new VueRouter({
    mode: 'history',
    routes: [...]
})
```

对比一下两种 URL 格式的差异：

hash 格式：http://localhost:8080/#/image

history 格式：http://localhost:8080/image

使用 history 格式后发现手动切换页面一切正常，但刷新页面时会提示页面不存在（404）。原因就是后端程序把 URL 解析了，而使用 hash 格式 URL 中会有一个 # 号分割，后端默认不会解析 # 后面的参数。

当使用 history 格式时，URL 就与正常的一样，例如 http://yoursite.com/user/id。不过这种格式需要后台配置支持。因为应用是个单页客户端应用，如果后台没有正确的配置，当用户在浏览器直接访问 http://oursite.com/user/id，就会返回 404。

所以要在服务端增加一个覆盖所有情况的候选资源：如果 URL 匹配不到任何静态资源，则应该返回同一个 index.html 页面，这个页面就是 app 依赖的页面。

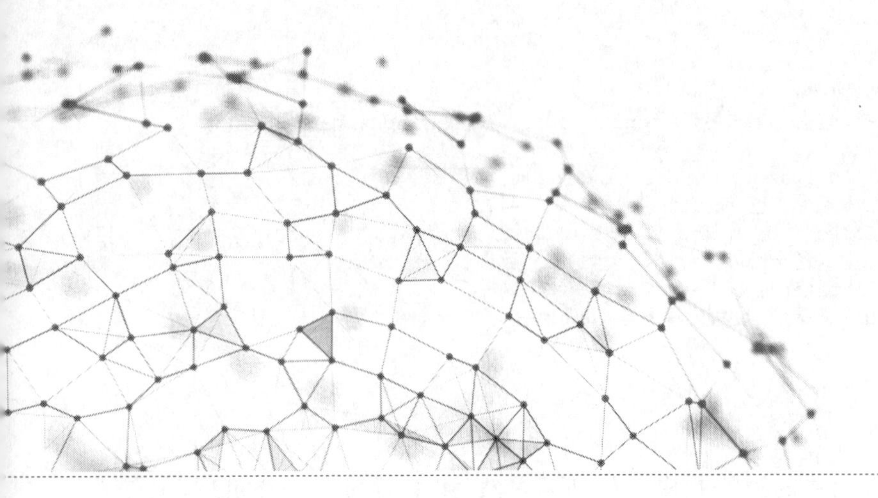

第12章 状态管理

在项目开发过程中，当组件比较多时，Vue 中各个组件之间传递数据是一件痛苦的事情，为此，Vuex 技术应运而生。Vuex 是一个状态管理的插件，可以解决不同组件之间的数据共享和数据持久化问题。使用 Vuex 来保存需要管理的状态值，值一旦被修改，所有引用该值的地方就会自动更新。

12.1 Vuex 概述

Vuex 是一个专为 Vue 应用程序开发的状态管理插件。它采用集中式存储管理应用的所有组件的状态，并以相应的规则保证状态以一种可预测的方式发生变化。Vuex 也集成到了 Vue 的官方调试工具 devtools extension，提供了诸如零配置的 time-travel 调试、状态快照导入导出等高级调试功能。

12.1.1 状态管理模式

状态管理模式其实就是集中存储管理应用的所有组件的状态。下面以一个简单的 Vue 计数应用说明：

```
new Vue({
  //state
  data () {
    return {
      count: 0
    }
  },
  //view
```

```
  template: '
    <div>{{ count }}</div>
  ',
  //actions
  methods: {
    increment () {
      this.count++
    }
  }
})
```

这个状态自管理应用包含以下 3 个部分。

- state：驱动应用的数据源。
- view：以声明方式将 state 映射到视图。
- actions：响应在 view 上的用户输入导致的状态变化。

图 12-1 是一个表示“单向数据流”理念的简单示意图。

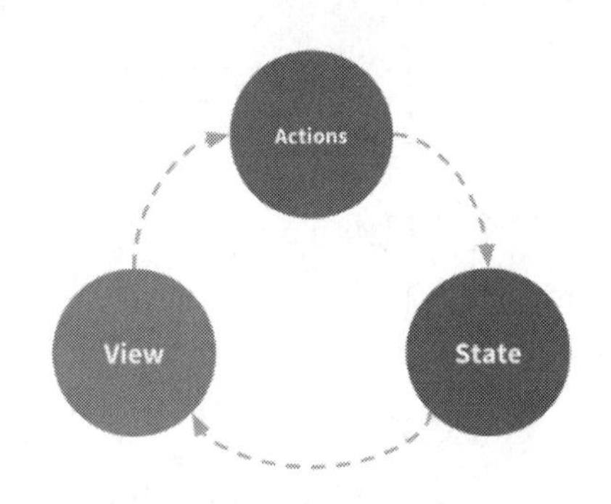

图 12-1 单向数据流

但是，当应用遇到多个组件共享状态时，单向数据流的简洁性很容易被破坏，出现以下两个问题：

（1）多个视图依赖于同一状态。

（2）来自不同视图的行为需要变更同一状态。

对于问题（1），多层嵌套的组件，传参的方法将会非常烦琐，并且对于兄弟组件间的状态传递无能为力。

对于问题（2），经常会采用父子组件直接引用或者通过事件来变更和同步状态的多份拷贝。

以上的这些模式非常脆弱，通常会产生无法维护的代码。因此，可以把组件的共享状态抽取出来，以一个全局单例模式管理，在这种模式下，组件树构成了一个巨大的“视图”，不管在树的哪个位置，任何组件都能获取状态或者触发行为。

通过定义和隔离状态管理中的各种概念并通过强制规则维持视图和状态间的独立性，代码将会变得更结构化且易维护。

这就是 Vuex 产生的背景，它借鉴了 Flux、Redux 和 The Elm Architecture。与其他模式不同的是，Vuex 是专门为 Vue 设计的状态管理库，以利用 Vue 的细粒度数据响应机制来进行高效的状态更新。

12.1.2 Vuex 的应用场合

Vuex 可以帮助我们管理共享状态，并附带了更多的概念和框架。这需要对短期和长期效益进行权衡。

如果不打算开发大型单页应用，使用 Vuex 可能是烦琐冗余的。如果您的应用够简单，最好不要使用 Vuex。一个简单的 store 模式就足够所需了。但是，如果需要构建一个中、大型单页应用，很可能会考虑如何更好地在组件外部管理状态，Vuex 将会成为合理的选择。

12.2 Vuex 的安装与使用

12.2.1 安装 Vuex

Vuex 的安装方法有 4 种，具体介绍如下。

（1）使用 CDN 引用，下载地址如下。

https://unpkg.com/vuex

unpkg.com 提供了基于 NPM 的 CDN 链接，它会一直指向 NPM 上发布的最新版本。用户也可以通过 https://unpkg.com/vuex@2.0.0 这样的方式指定特定的版本。

（2）在 Vue 之后引入 vuex 包进行安装，代码如下：

```
<script src="/path/to/vue.js"></script>
<script src="/path/to/vuex.js"></script>
```

（3）使用 NPM 安装，代码如下：

```
npm install vuex --save
```

（4）使用 Yarn 安装，代码如下：

```
yarn add vuex
```

注意

在一个模块化的打包系统中，必须显式地通过 Vue.use() 来安装 Vuex：

```
import Vue from 'vue'
import Vuex from 'vuex'
Vue.use(Vuex)
```

当使用全局 script 标签引用 Vuex 时，不需要以上安装过程。

12.2.2 Promise 对象

Vuex 依赖 Promise。Promise 是一个对象，它代表了一个异步操作的最终完成或者失败。

如果支持的浏览器并没有实现 Promise(例如 IE)，可以使用一个 polyfill 的库，例如 es6-promise，可以通过 CDN 将其引入：

```
<script src="https://cdn.jsdelivr.net/npm/es6-promise@4/dist/es6-promise.auto.
js"></script>
```

引入后 window.Promise 会自动可用。

如果使用 NPM 或 Yarn 等包管理器，可以按照下列方式执行安装：

```
npm install es6-promise --save # npm
yarn add es6-promise # Yarn
```

也可以将下列代码添加到使用 Vuex 之前的一个地方进行安装：

```
import 'es6-promise/auto'
```

12.2.3 使用 Vuex

Vuex 应用的核心就是 store（仓库）。store 基本上就是一个容器，它包含着应用中大部分的状态 (state)。Vuex 和单纯的全局对象有以下两点不同。

（1）Vuex 的状态存储是响应式的。当 Vue 组件从 store 中读取状态的时候，若 store 中的状态发生变化，那么相应的组件也会相应地得到高效更新。

（2）不能直接改变 store 中的状态。改变 store 中的状态的唯一途径就是显式地提交 (commit)mutation。这样可以方便地跟踪每一个状态的变化。

先来看一个计数的示例，单击按钮可以增、减 message 的值。

```
<div id="app">
    <p>计数</p>
    <button @click="reduce">减少</button>
    <button>{{message}}</button>
    <button @click="add">增加</button>
</div>
<script>
    new Vue({
        el:"#app",
        data:{
            message:"0"
        },
        methods:{
            add:function () {
                this.message++
            },
            reduce:function () {
                this.message--
```

```
            }
        },
    })
</script>
```

安装 Vuex 之后，就可以使用 store 来管理示例中的数据了，仅需要提供一个初始 state 对象和一些 mutation。下面是更改后的完整代码：

```
<div id="app">
    <p>计数</p>
    <button @click="reduce">减少</button>
    <button>{{message}}</button>
    <button @click="add">增加</button>
</div>
<script>
    //创建store
    const store=new Vuex.Store({
        //初始 state 对象，存储message
        state:{
            message:0
        },
        //mutations对象用来更改状态，也是唯一一个可以更改状态的地方
        mutations:{
            //定义更改状态的方法
            addMsg:function(state){
                state.message++;    //更改state中message（增加）
            },
            reduceMsg:function(state){
                state.message--;  //更改state中message（减少）
            }
        }
    })
    new Vue({
        el:"#app",
        store,  //在vue对象中注册store
        //使用计算属性（computed）关联message
        computed:{
            message:function(){
                //获取仓库中的message数据
                return this.$store.state.message;
            }
        },
        methods:{
            add:function () {
                //关联状态管理中的方法addMsg，使用commit触发状态变更
                store.commit('addMsg')
            },
            reduce:function () {
                //关联状态管理中的方法reduceMsg，使用commit触发状态变更
                store.commit('reduceMsg')
            }
        },
    })
</script>
```

在谷歌浏览器中运行，并单击“增加”按钮，页面效果如图 12-2 所示。

图 12-2　页面效果

从上面的示例可以知道：由于 store 中的状态是响应式的，在组件中调用 store 中的状态简单到仅需要在计算属性中返回即可，触发变化也仅仅是在组件的 methods 中提交 mutation。

提示

可以发现更改后的代码，比不使用 Vuex 代码量变大，所以在使用 Vuex 的时候要根据自己的项目去选择。Vuex 多用于中、大型项目，而对于小型项目，推荐使用 HTML 5 特有的属性 localStroage 和 sessionStroage 作为数据之间的传递。

注意

如果在模块化构建系统中，请确保在开头调用了 Vue.use(Vuex)。

12.3 在项目中使用 Vuex

12.3.1　使用脚手架搭建一个项目

下面使用脚手架来搭建一个排行榜的项目，在命令行中输入并执行 vue create billboard 命令，在接下来配置选项中，直接使用默认设置即可。具体的操作请参考“脚手架”一章。

项目创建完成后，目录结构如图 12-3 所示，根据需求把旧结构更改成如图 12-4 所示的目录。

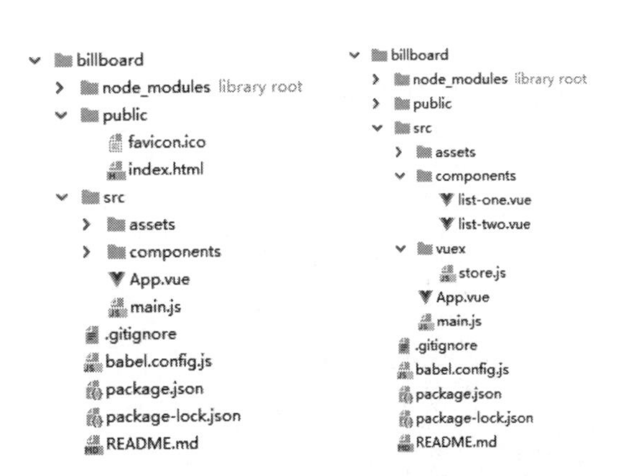

图 12-3　目录结构　　图 12-4　更改后的目录结构

从目录结构上看，在components文件夹中添加了list-one和list-two两个组件，在src文件夹中添加了vuex状态管理文件夹。对于list-one和list-two组件，需要在App.vue中引入，App.vue组件代码如下：

```
<template>
  <div id="app">
    <list-one></list-one>
    <list-two></list-two>
  </div>
</template>
<script>
  //引入组件
import ListOne from './components/list-one.vue'
import ListTwo from './components/list-two.vue'
export default {
  name: 'app',
  components: {
    //注册组件
    'list-one':ListOne,
    'list-two':ListTwo
  }
}
</script>
```

使用Vuex之前，先在项目中安装Vuex状态管理，在命令行中执行npm install vuex --save命令。安装完成后，store.js文件内容的配置如下：

```
//先引入vue和vuex
import Vue from 'vue'
import Vuex from 'vuex'
//然后通过全局方法Vue.use()使用Vuex
Vue.use(Vuex);
//创建store实例
const store=new Vuex.Store({
    //内容
})
//导出store
export default store;
```

内容配置成后，还需要在程序入口文件main.js文件中引入并注册，main.js代码如下：

```
import Vue from 'vue'
import App from './App.vue'
import store from './vuex/store.js'  //引入
Vue.config.productionTip = false
new Vue({
  store,          // 注册
  render: h => h(App),
}).$mount('#app')
```

到这里，整个项目配置完成，也引入了Vuex状态管理。

下面来创建一个“前端框架排行榜”的小项目，在list-one和list-two两个组件中编写内容。

```
<template>
  <div id="app">
    <!--在组件中绑定frames-->
    <list-one v-bind:frames="frames"></list-one>
    <list-two v-bind:frames="frames"></list-two>
  </div>
</template>
<script>
  //引入组件
import ListOne from './components/list-one.vue'
import ListTwo from './components/list-two.vue'
export default {
  name: 'app',
  components: {
    //注册组件
    'list-one':ListOne,
    'list-two':ListTwo
  },
  data(){
    return {
      frames: [
        {name: 'Vue.js', star: 15},
        {name: 'React.js', star: 13},
        {name: 'Angular.js', star: 12},
      ]
    }
  }
}
</script>
```

list-one 组件代码如下：

```
<template>
    <div id="list-one">
        <h2>框架排行（组件one）</h2>
        <ul>
            <li v-for="frame in frames">
                <span class="name">{{frame.name}}——</span>
                <span class="star">热度：{{frame.star}}</span>
            </li>
        </ul>
    </div>
</template>
<script>
    export default {
        name: "list-one",
        props:["frames"]  //接收父组件传过来的值
    }
</script>
<style scoped>
    #list-one{
        background: #ffec44;
        box-shadow: 1px 2px 3px rgba(0,0,0,0.2);
        margin-bottom: 30px;
        padding: 10px 20px;
    }
    #list-one ul{
        padding: 0;
    }
```

```
    #list-one li{
        display: inline-block;
        margin-right: 10px;
        margin-top: 10px;
        padding: 20px;
        background: rgba(255,255,255,0.7);
    }
    .star{
        font-weight: bold;
        color: #e829cc;
    }
</style>
```

list-two 组件代码如下：

```
<template>
    <div id="list-two">
        <h2>框架排名（组件two）</h2>
        <ul>
            <li v-for="frame in frames">
                <span class="name">{{frame.name}}</span>
                <span class="star">热度：{{frame.star}}</span>
            </li>
        </ul>
    </div>
</template>
<script>
    export default {
        name: "list-two",
        props:["frames"]  //接收父组件传过来的值
    }
</script>
<style scoped>
    #list-two{
        background: #b7fcff;
        box-shadow: 1px 2px 3px rgba(0,0,0,0.2);
        margin-bottom: 30px;
        padding: 10px 20px;
    }
    #list-two ul{
        padding: 0;
        list-style-type: none;
    }
    #list-two li{
        margin-right: 10px;
        margin-top: 10px;
        padding: 20px;
        background: rgba(255,255,255,0.7);
    }
    .star{
        font-weight: bold;
        color: #2618e8;
        display: block;
    }
</style>
```

编写完成后，执行 npm run serve 命令运行项目，在谷歌浏览器中输入“http://localhost:8080/”并打开，页面效果如图 12-5 所示。

图 12-5　项目运行效果

注意

在使用 props 访问父组件数据时，需要在根组件中使用 v-bind 绑定数据。

```
<list-one v-bind:frames="frames"></list-one>
<list-two v-bind:frames="frames"></list-two>
```

这个项目中，使用 Vue 的属性把数据传递给不同的组件。在接下来的内容中，将使用 Vuex 来实现数据的管理。

12.3.2　state 对象

在上面的项目中，可以把共用的数据提取出来，放到状态管理的 state 对象中。

```
state:{
        frames: [
            {name: 'Vue.js', star: 15},
            {name: 'React.js', star: 13},
            {name: 'Angular.js', star:12},
        ]
}
```

由于 Vuex 的状态存储是响应式的，从 store 实例中读取状态最简单的方法就是在计算属性中返回某个状态。

组件 list-one 的代码如下：

```
export default {
        name: "list-one",
        computed:{
            frames:function(){
                return this.$store.state.frames
            }
        }
    }
```

组件 list-two 的代码如下：

```
export default {
        name: "list-two",
        computed:{
            frames:function(){
                return this.$store.state.frames
            }
        }
    }
```

更改后重新执行 npm run serve 命令运行项目，组件即可获取到数据。但是组件中每一个属性都是函数，如果有许多个，那么就要写很多函数，且需要重复写 return this.$store.state，有些重复和冗余。Vue 提供了 mapState 辅助函数，它把 state 直接映射到我们的组件中。

使用 mapState 辅助函数前，需要先引入 mapState 辅助函数，代码如下：

```
import {mapState} from "vuex"; //引入mapState
```

把组件中的组件 list-one 和组件 list-two 中计算属性换成如下代码：

```
computed:mapState({
frames:'frames'//'frames'直接映射到state对象中的frames，它相当于this.$store.state.
frames
})
```

重新运行项目，可发现其运行结果和上面的示例一样。

上面是使用对象的方法，还可以使用数组的方法获取数据，代码如下：

```
computed:mapState([
    'frames'
])
```

12.3.3 getter 对象

有时候组件中获取到 store 中的 state 后，需要对其进行加工才能使用，computed 属性中就需要写操作函数，如果有多个组件中都需要进行这个操作，那么在各个组件中都要写相同的函数，非常麻烦。这时可以把这个相同的操作写到 store 中的 getters 对象中。每个组件只要引用 getter 就可以了，非常方便。

getter 就是把组件中共有的对 state 的操作进行了提取，它就相当于是 state 的计算属性。getter 的返回值会根据它的依赖被缓存起来，且只有当它的依赖值发生了改变才会被重新计算。

提示

getter 接受 state 作为其第一个参数。

例如，在上面项目的两个组件中，为排行榜中的每个 name 值都添加“*”号。

首先在 getters 对象中，添加操作的方法：

```
getters:{
        //定义方法varyFrames，传入state
        varyFrames:function(state){
            //使用map方法循环遍历frames
            var varyFrames=state.frames.map(frames=>{
                    return {
                        //对frames进行操作
                        name:"**"+frames.name+"**",
                        star:frames.star
                    }
                }
            )
        return varyFrames;
        }
    }
```

方法定义完成以后，便可以在想要使用的组件中使用。在组件的计算属性中进行引入。在 list-one 和 list-two 组件中分别引入：

```
computed:{
            frames:function(){
                return this.$store.state.frames
            },
            varyFrames:function () {
                return this.$store.getters.varyFrames;
    }
}
```

并把循环遍历的 frames 换成 varyFrames，因为 varyFrames 是变更后的数组。

```
<li v-for="frame in varyFrames">
```

然后重新运行项目，页面效果如图 12-6 所示。

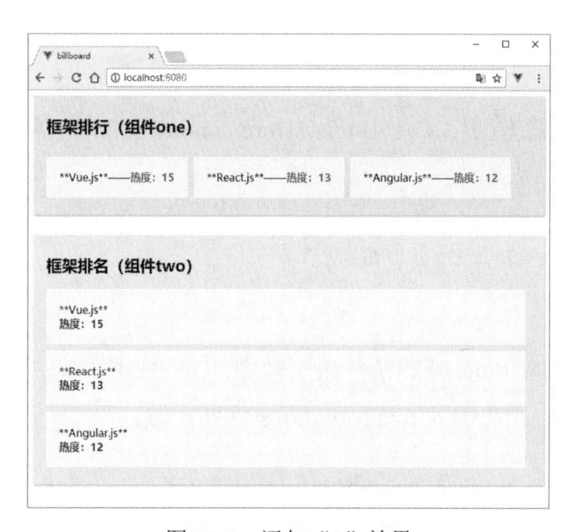

图 12-6　添加“*”效果

和 state 对象一样，getters 对象也有一个辅助函数 mapGetters，它可将 store 中的 getter 映射到局部计算属性中，使用前应先引入辅助函数 mapGetters：

```
import { mapGetters } from 'vuex'
```

上面代码可简化，代码如下：

```
...mapGetters([
      'varyFrames'
])
```

如果想将一个 getter 属性另取一个名字，可使用对象形式，代码如下：

```
...mapGetters({
     varyFramesOne:'varyFrames'
})
```

12.3.4 mutation 对象

更改 Vuex 的 store 中的状态，唯一方法就是提交 mutation。Vuex 中的 mutation 类似于事件。每个 mutation 都有一个字符串的事件类型 (type) 和一个回调函数 (handler)。这个回调函数就是实际进行状态更改的地方，并且它会接受 state 作为第一个参数。

下面在项目中的 list-one 中添加一个 <button> 按钮，当单击这个按钮时，项目中所有组件的 star 的数量都将增加，这时就需要在 mutation 中进行定义。更改的数据将会渲染到所有组件中。

下面是 mutations 中定义的方法：

```
mutations:{
    addStar:function(state){
        state.frames.map(function(frames){
            frames.star+=1;
        })
    }
}
```

在组件 list-one 中的方法中，使用 this.$store.commit'addStar' 来提交 mutation：

```
methods:{
    addStar:function(){
        this.$store.commit('addStar')
    }
}
```

也可以使用 mapMutations 辅助函数将组件中的 methods 映射为 store.commit 调用（需要在根节点注入 store）。先引入 mapMutations 辅助函数：

```
import {mapMutations} from "vuex"
methods:{
            //数组形式
```

```
            ...mapMutations([
//将 'this.addStar()' 映射为 'this.$store.commit('addStar')'
        'addStar',
            ]),
            //对象形式
            ...mapMutations({
                //更名使用对象方式
                add:'addStar',
            }),
        }
```

在 list-one 组件中对 star 进行更改后，所有组件中的 star 数据都会发生改变。下面重新运行项目，然后单击“增加热度”按钮，可以发现组件 lsit-two 中的数据也发生了改变，如图 12-7 所示。

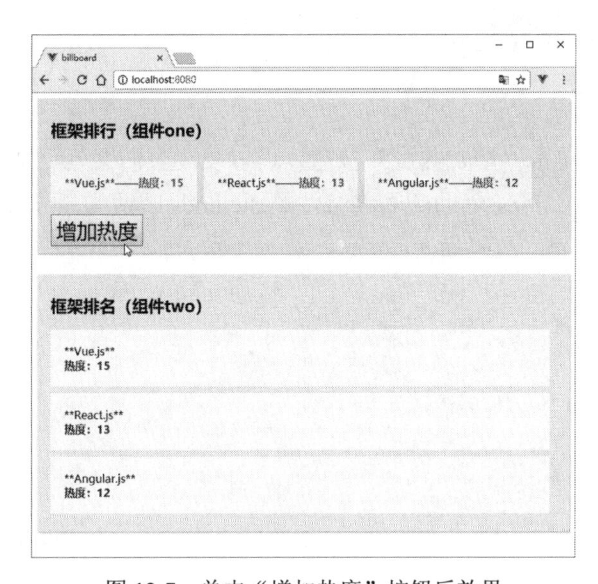

图 12-7　单击“增加热度”按钮后效果

12.3.5　Action 对象

Action 类似于 mutation，不同点在于以下两个方面：

- Action 提交的是 mutation，而不是直接变更状态。
- Action 可以包含任意异步操作。

在 Vuex 中提交 mutation 是更改状态的唯一方法，并且这个过程是同步的，异步逻辑都应该封装到 action 里面。

Action 函数接受一个与 store 实例具有相同方法和属性的 context 对象，因此可以调用 context.commit 提交一个 mutation，或者通过 context.state 和 context.getters 来获取 state 和 getters。

在上面项目中，使用 Action 对象执行异步操作，单击按钮后延迟 2 秒，增加 star 的数量，代码如下。

```
actions:{
        //context对象类似于store 实例
        addStar:(context)=>{
            //2秒后执行方法
            setTimeout(function(){
                //激活addStar方法
                context.commit('addStar')
            },2000);
        }
    }
```

在 lsit-one 组件的 methods 中触发：

```
methods:{
    addStar:function(){
        //Action对象通过store.dispatch方法触发
        this.$store.dispatch('addStar')
    }
}
```

重启项目，然后单击“增加热度”按钮，可以发现页面会延迟 2 秒后再加 1。

还可以使用 mapActions 辅助函数将组件的 methods 映射为 store.dispatch 调用（需要先在根节点注入 store），代码如下：

```
import { mapActions } from 'vuex'
methods:{
    ...mapActions([
        'addStar'
    ]),
    ...mapActions({
        //更名使用对象形式
        add:'addStar'
    })
}
```

在 Action 对象中，还可以传递参数，例如，单击一次按钮，让 star 增加 5。更改项目代码如下。

（1）组件 List-one。

```
<button @click="addStar(5)">增加热度</button>
methods:{
        addStar:function(count){
            //Action 通过 store.dispatch 方法触发
            this.$store.dispatch('addStar',count)
        }
    }
```

（2）Store.js。

```
mutations:{
        addStar:function(state,payload){
            state.frames.map(function(frames){
                frames.star+=payload;
            })
        }
```

```
    },
    actions:{
        //context类似于store 实例
        addStar:(context,payload)=>{
            //2秒后执行方法
            setTimeout(function(){
                //激活addStar方法
                context.commit('addStar',payload)
            },2000);
        }
    }
```

在谷歌浏览器中运行，单击按钮后，可以发现延迟 2 秒后，star 直接增加 5。

12.4 疑难解惑

疑问 1：在 Vuex 中，如何开启严格模式？在严格模式下有什么效果？

开启严格模式，仅需在创建 store 的时候传入 strict:true，代码如下。

```
const store = new Vuex.Store({
    //...
    strict: true
})
```

在严格模式下，无论何时发生了状态变更且不是由 mutation 函数引起的，都将会抛出错误。这能保证所有的状态变更都能被调试工具跟踪到。

疑问 2：在 Vuex 中如何使用插件？

Vuex 的 store 接 plugins 选项，这个选项暴露出每次 mutation 的钩子。Vuex 插件就是一个函数，它接受 store 作为唯一参数，下面定义 myPlugin 插件：

```
const myPlugin = store => {
  //当 store 初始化后调用
  store.subscribe((mutation, state) => {
    //每次 mutation 之后调用
    //mutation 的格式为 { type, payload }
  })
}
```

使用 myPlugin 插件：

```
const store = new Vuex.Store({
  // ...
  plugins: [myPlugin]
})
```

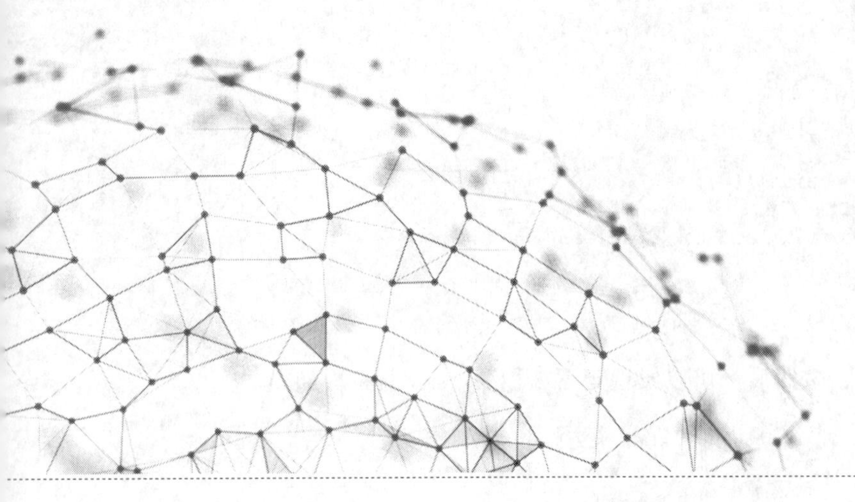

第13章 项目实训1——神影视频App

本章将介绍一个电影网站App——神影视频，它使用Vue脚手架进行搭建，页面简洁、精致，和一些常见的电影购票网站类似，例如支付宝中的“淘票票电影”。

13.1 准备工作

13.1.1 开发环境

本项目的开发环境如下：

（1）编辑器 Webstrom。

（2）Node 10.16.0。

（3）NPM 6.9.0。

（4）Vue 脚手架 3.8.3。

（5）测试浏览器 Google（版本 59.0.3071.104，开发者内部版本）。

13.1.2 搭建 Vue 脚手架

本项目使用 Vue 脚手架进行搭建，关于脚手架的全局安装请参考本书“脚手架”章节。下面介绍一下项目的创建过程。

（1）首先打开项目目录（要创建项目的目录），如图 13-1 所示。

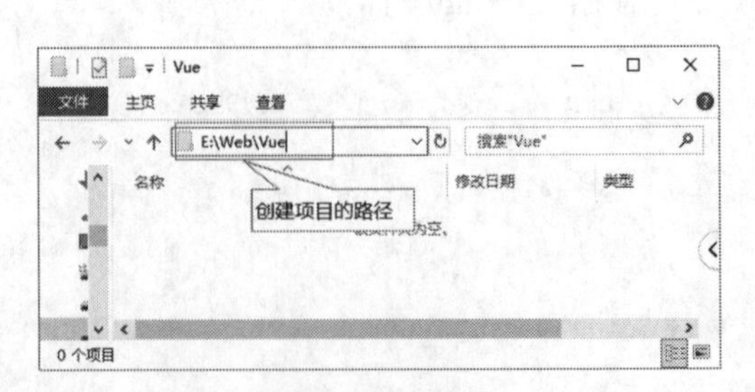

图 13-1　要创建项目的目录

（2）打开命令提示符窗口，先切换对应的磁盘，然后在窗口中输入 cd E:\Web\Vue 命令，按 Enter 键进入项目路径，如图 13-2 所示。

（3）vue create 语句创建项目，项目名称为“shenying”，注意项目的名称不能大写，否则无法创建，如图 13-3 所示。

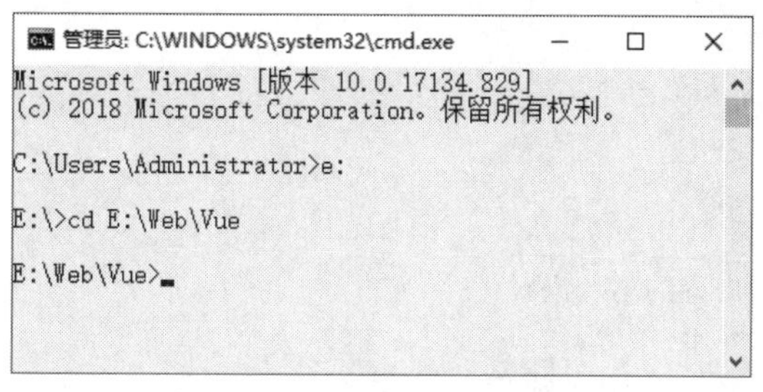

图 13-2　进入项目路径

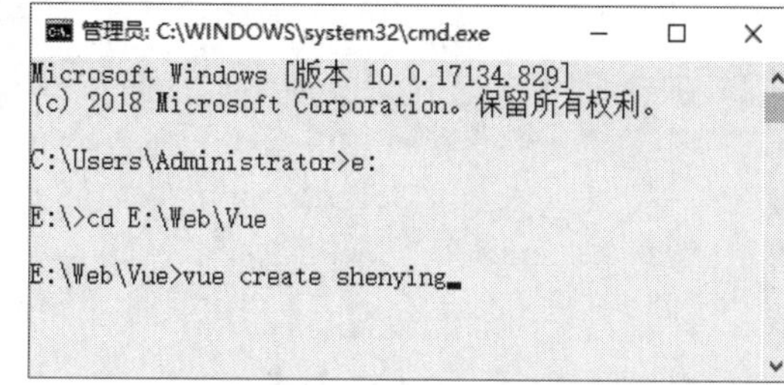

图 13-3　创建 shenying 项目

（4）按 Enter 键后，会提示两种项目的配置选项，如图 13-4 所示。

这里选择自定义的配置，根据项目需要，选择合适的模块，然后按 Enter 键即可。本项目选择的模块如图 13-5 所示。

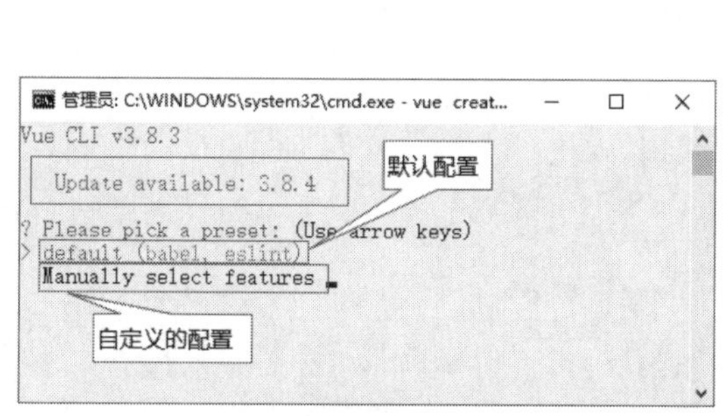

图 13-4　选择配置方式

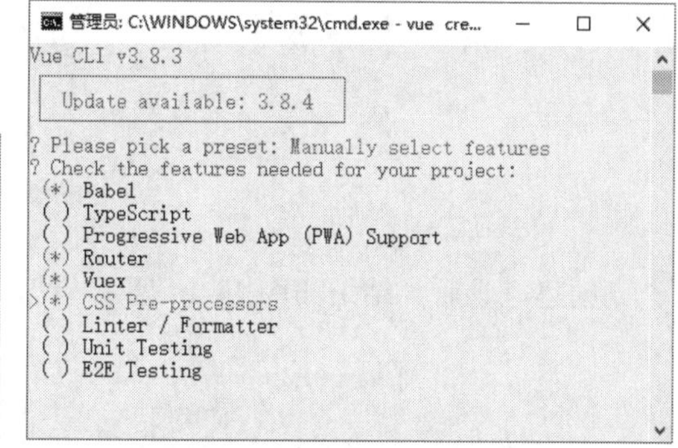

图 13-5　项目模块选择

模块选择完成后，按 Enter 键，弹出提示，询问是否使用 history 模式的路由，这里选择使用，在命令行输入“y”，如图 13-6 所示，然后按 Enter 键。

接下来，又提示选择预编译的 CSS，这里选择 Sass/SCSS (with dart-sass)，如图 13-7 所示。

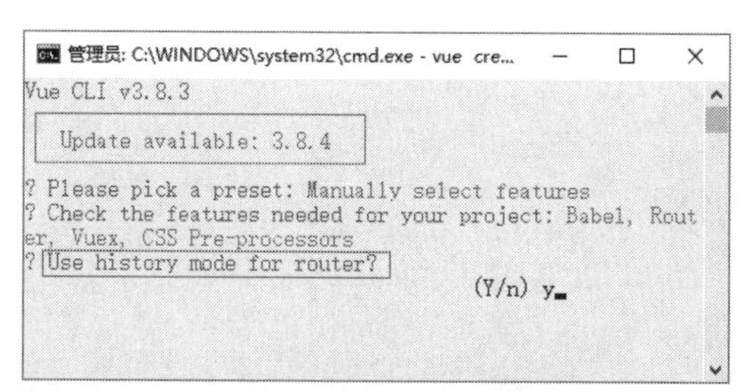

图 13-6　是否使用 history 模式的路由

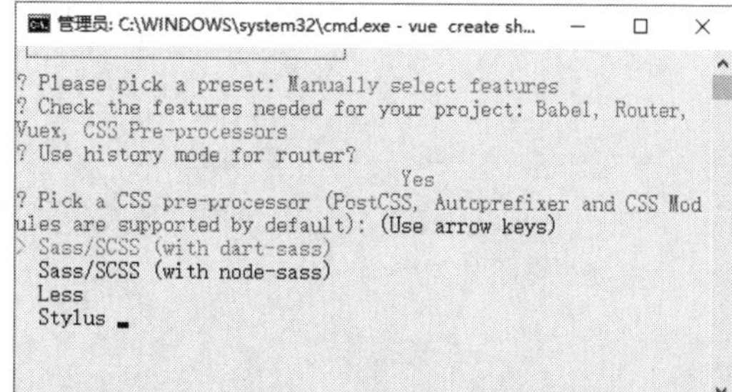

图 13-7　选择预编译的 CSS

然后，提示配置内容写入的位置，选择“In package.json”，写入到package.json文件中，如图13-8所示。

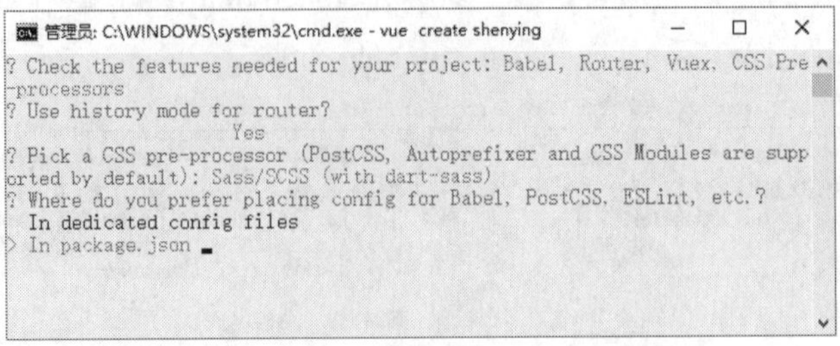

```
管理员: C:\WINDOWS\system32\cmd.exe - vue  create shenying
? Check the features needed for your project: Babel, Router, Vuex, CSS Pre-processors
? Use history mode for router?
                    Yes
? Pick a CSS pre-processor (PostCSS, Autoprefixer and CSS Modules are supported by default): Sass/SCSS (with dart-sass)
? Where do you prefer placing config for Babel, PostCSS, ESLint, etc.?
  In dedicated config files
> In package.json
```

图13-8　选择配置内容的写入位置

最后提示，是否把上面配置的内容作为将来项目的预置，这里不进行设置，在命令行输入“n”，按Enter键。

（5）到这里选项已经配置完成，接下来就是项目的创建，如图13-9所示。

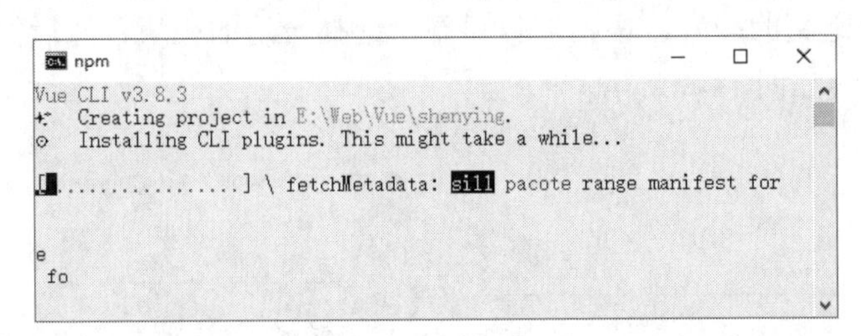

图13-9　开始创建项目

创建完成后，结果如图13-10所示。

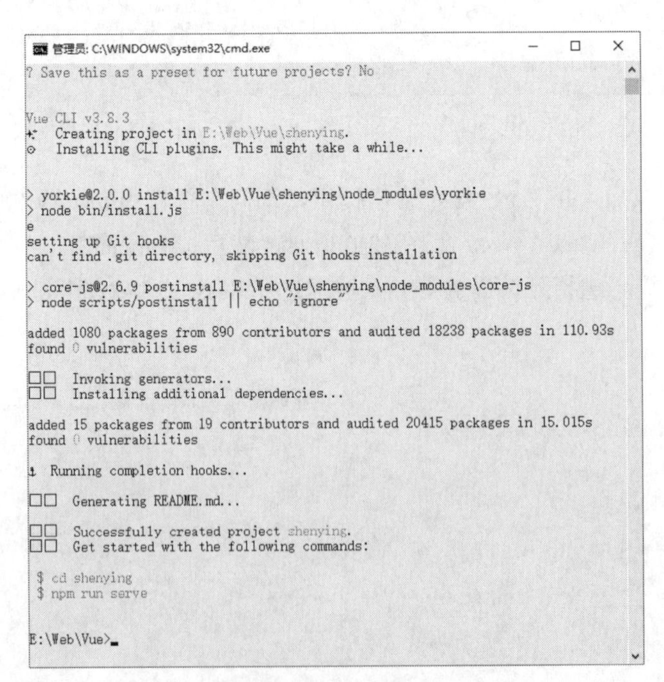

图13-10　完成创建

打开项目目录结构，如图 13-11 所示。

图 13-11　项目目录结构

项目搭建完成后，测试一下是否能正常启动。根据提示，首先进入到项目目录，在命令行中输入“cd shenying”，按 Enter 键；然后再输入“npm run serve”启动项目。

项目启动成功之后，提供两个域名，供我们测试使用，如图 13-12 所示。

在谷歌浏览器中输入本地域名“http://localhost:8080/”，按 Enter 键。打开谷歌浏览器的控制台，选择使用谷歌浏览器的模拟手机效果，如图 13-13 所示。

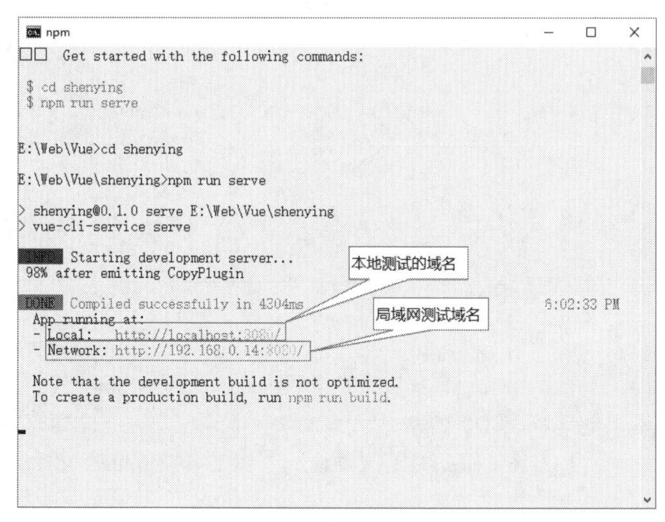

图 13-12　测试域名

图 13-13　项目运行效果

13.2 网站概述

本项目使用的都是本地静态的资源，主要是前端展示，没有涉及后台的开发。

13.2.1　网站结构

使用脚手架搭建的项目，目录结构可以根据自己的喜好进行更改，但是要注意进行

相应的配置。

其中，public 文件夹用来放置项目的静态文件，结构如图 13-14 所示。

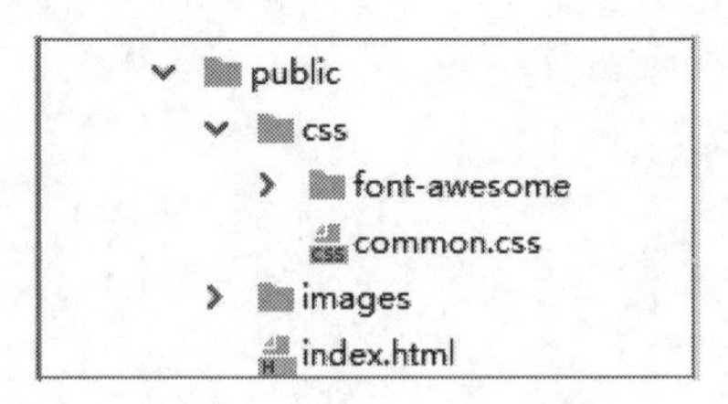

图 13-14　public 文件夹结构

在 src 文件夹中，放置了所有的源码文件。其中 components 文件夹用来放置一些比较小的、公用的组件，例如，头部和尾部组件，结构如图 13-15 所示；views 文件夹用来放置 3 个主页面的组件，是页面级别的组件，如图 13-16 所示；routers 文件夹用来放置路由，其中 index.js 文件是主路由，目录结构如图 13-17 所示；main.js 是项目入口文件的 JavaScript 逻辑，在 webpack 打包之后将被注入 index.html 页面中。

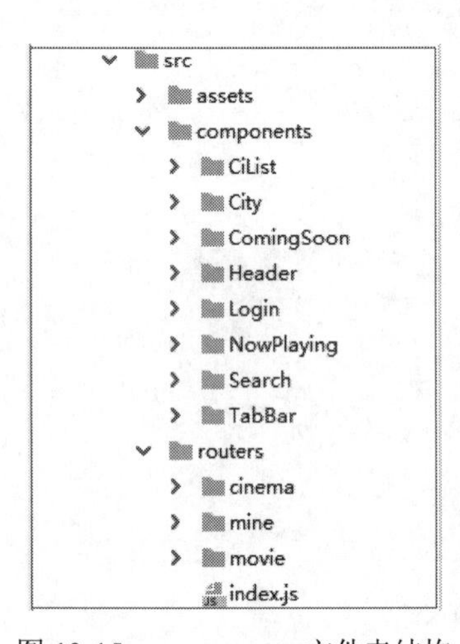

图 13-15　components 文件夹结构

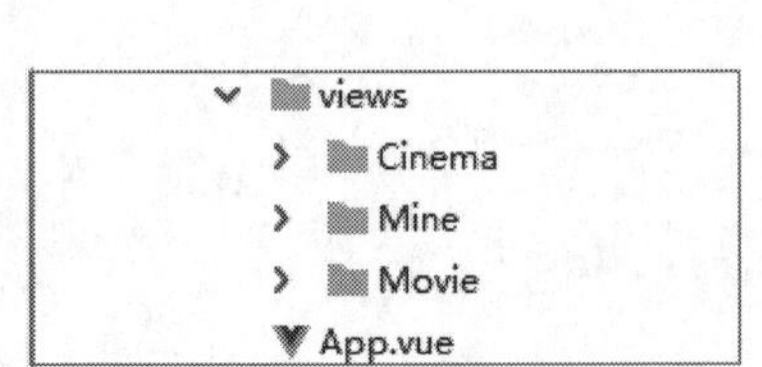

图 13-16　views 文件夹结构

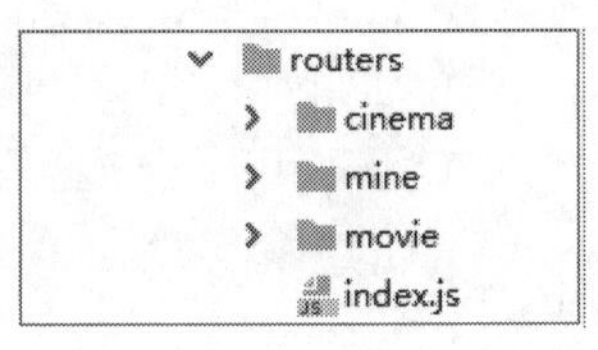

图 13-17　routers 文件夹结构

13.2.2　初始化项目文件

本项目更改了项目的文件，所以需要更改一些引用或配置。

1. 主组件（app.vue）

```
<template>
<!--  <keep-alive>是Vue的内置组件，能在组件切换过程中将状态保留在内存中，防止重复渲染DOM。-->
    <keep-alive>
        <router-view></router-view>
    </keep-alive>
</template>
```

2. 主模块（main.js）

```
import Vue from 'vue'
import App from './App.vue'
import router from './routers'
import store from './stores'
Vue.config.productionTip = false
new Vue({
  router,
  store,
  render: h => h(App)
}).$mount('#app')
```

3. 主页面（index.html）

在 public 目录下的 index.html 文件，是脚手架搭建的项目的默认主页面，Vue 的所有组件都渲染到 <div id="app"></div> 中。

```
<!DOCTYPE html>
<html lang="en">
  <head>
    <meta charset="utf-8">
    <meta http-equiv="X-UA-Compatible" content="IE=edge">
    <!-- user-scalable=no 不让用户进行缩放-->
     <meta name="viewport" content="width=device-width,initial-scale=1.0,user-
scalable=no">
    <!--<%= BASE_URL %>在路由中定义-->
    <link rel="icon" href="<%= BASE_URL %>favicon.ico">
    <link rel="stylesheet" href="<%= BASE_URL %>css/common.css">
     <link rel="stylesheet" href="<%= BASE_URL %>css/font-awesome/css/font-
awesome.css">
    <title>shenying</title>
  </head>
  <body>
    <noscript>
      <strong>We're sorry but shenying doesn't work properly without JavaScript
enabled. Please enable it to continue.</strong>
    </noscript>
    <!-- 所有的组件都会渲染到app元素中-->
    <div id="app"></div>
    <!-- built files will be auto injected -->
  </body>
</html>
```

其中，引入的 common.css 文件是项目的公用样式，font-awesome.css 是字体样式。

13.3 设计项目组件

组件的设计是把重复利用的内容进行组件化，方便调用。本项目组件是在 components 目录中进行定义。

13.3.1 设计头部和底部导航组件

本项目主要由 3 个页面构成，分别是电影页面、影院页面和我的页面。三个页面都包括头部内容和底部的导航栏，可以分别设计成组件，在每个页面中引入即可。

1. 头部组件（Header）

```
<template>
    <header id="header">
        <h1>{{title}}</h1>
    </header>
</template>
<script>
    export default {
        name: "Header",
        //props是子组件访问父组件数据的唯一接口
        props:{
            title:{
                type:String,
                default:'神影电影'
            }
        }
    }
</script>
<!--scoped属性实现了私有化的样式-->
<style scoped>
    #header{
        width: 100%;
        height: 50px;
        color: #ffffff;
        background: #e54847;
        border-bottom:1px solid #e54847;
        position: relative;
    }
    #header h1{
        font-size: 18px;
        text-align: center;
        line-height: 50px;
        font-weight: normal;
    }
    #header i{
        position: absolute;
        left: 5px;top: 50%;
        margin-top: -13px;
        font-size: 26px;
    }
</style>
```

头部组件在谷歌浏览器中运行效果如图 13-18 所示。

神影电影

图 13-18　头部组件效果

2. 底部导航组件（TabBar）

在底部的导航中，配置一级路由，用来切换主页面组件：电影页面、影院页面和我的页面。文件中，<router-link> 组件支持用户在具有路由功能的应用中单击导航。通过 to 属性指定目标地址，默认渲染为带有正确连接的 <a> 标签，可以通过配置 tag 属性生成别的标签。另外，当目标路由成功激活时，链接元素自动设置一个表示激活的 CSS 类名，代码如下：

```
<template>
    <div id="footer">
        <ul>
            <router-link tag="li" to="/movie">
                <i class="fa fa-film"></i>
                <p>电影</p>
            </router-link>
            <router-link tag="li" to="/cinema">
                <i class="fa fa-youtube-square"></i>
                <p>影院</p>
            </router-link>
            <router-link tag="li" to="/mine">
                <i class="fa fa-user-circle"></i>
                <p>我的</p>
            </router-link>
        </ul>
    </div>
</template>
<script>
    export default {
        name: "Tabbar"
    }
</script>
<style scoped>
    #footer{
        width: 100%;
        height: 50px;
        background: white;
        border-top: 2px solid #ebe8e3;
        position: fixed;
        left: 0;
        bottom: 0;
    }
    #footer ul{
        display: flex;
        text-align: center;
        height: 50px;
        align-items: center;
    }
    #footer ul li{
        flex: 1;
        height: 40px;
    }
    #footer li.active{color: #f03d37;}
    /* router-link-active:路由中自带的样式，选中时的颜色*/
    #footer li.router-link-active{color: #f03d37;}
    #footer ul i{font-size: 20px;}
    #footer ul p{
        font-size: 12px;
        line-height: 18px;
    }
</style>
```

配置路由 index.js 文件，代码如下：

```
import Vue from 'vue'
import Router from 'vue-router'
import movieRoter from './movie'
import cinemaRoter from './cinema'
import mineRoter from './mine'
//./router 代表 touter 文件夹里的 index文件，其他文件要加名字
Vue.use(Router)
//export defalut 是输出，相当于把接口暴露在外部，供所有文件来调用
export default new Router({
    //mode可选参数
    //hash: 默认为hash，如果使用hash的话，页面的地址就会加上 # 号，会比较不好看，如我
    //们的地址变成：http://localhost:8080/#/
    //history: 我们使用history的话，那么访问页面的时候就和平常一样，不带#号
    mode: 'history',
    base: process.env.BASE_URL,
    routes: [
        movieRoter,
        cinemaRoter,
        mineRoter,
        //重定向：当上面路由都不匹配的情况下，跳转到movie页面
        {
            path:'/*',
            redirect:'/movie'
        }
    ]
})
```

底部导航组件在谷歌浏览器中的运行效果如图 13-19 所示。

图 13-19　底部导航组件效果

13.3.2　设计电影页面组件

电影页面有 4 个组件，分别为：城市、正在热映、即将上映和搜索。

1. 城市组件（City）

在城市组件中，只列举了首字母以 A、B、C、D、E 开头的城市。

```
<template>
    <div class="city_body">
        <div class="city_list">
            <div class="city_hot">
                <h2>热门城市</h2>
                <ul class="clearfix">
                    <li>北京</li>
                    <li>上海</li>
                    <li>天津</li>
                    <li>合肥</li>
```

```
                    <li>郑州</li>
                </ul>
            </div>
            <div class="city_sort">
                <div>
                    <h2>A</h2>
                    <ul>
                        <li>阿克苏</li>
                        <li>安康</li>
                        <li>安庆</li>
                    </ul>
                </div>
                <div>
                    <h2>B</h2>
                    <ul>
                        <li>白山</li>
                        <li>白城</li>
                        <li>宝鸡</li>
                    </ul>
                </div>
                <div>
                    <h2>C</h2>
                    <ul>
                        <li>沧州</li>
                        <li>长春</li>
                        <li>昌吉</li>
                    </ul>
                </div>
                <div>
                    <h2>D</h2>
                    <ul>
                        <li>大理</li>
                        <li>大连</li>
                        <li>大庆</li>
                    </ul>
                </div>
                <div>
                    <h2>E</h2>
                    <ul>
                        <li>鄂尔多斯</li>
                        <li>恩施</li>
                        <li>鄂州</li>
                    </ul>
                </div>
            </div>
        </div>
        <div class="city_index">
            <ul>
                <li>A</li>
                <li>B</li>
                <li>C</li>
                <li>D</li>
                <li>E</li>
            </ul>
        </div>
    </div>
</template>
<script>
    export default {
```

```
            name: "City"
        }
    </script>
    <style scoped>
        #content .city_body{
            margin-top: 45px;
            display: flex;
            width: 100%;
            position: absolute;
            top: 0;
            bottom: 0;
        }
        .city_body .city_list{
            flex: 1;
            overflow: auto;
            background: #fff5f0;
        }
        .city_body .city_list::-webkit-scrollbar{
            background-color: transparent;
            width: 0;
        }
        .city_body .city_hot{
            margin-top: 20px;
        }
        .city_body .city_hot h2{
            padding-left: 15px;
            line-height: 30px;
            font-size: 14px;
            background: #f0f0f0;
            font-weight:normal;
        }
        .city_body .city_hot ul li{
            float: left;
            background: #fff;
            width: 29%;
            height: 33px;
            margin-top: 15px;
            margin-left: 3%;
            padding:0 4px;
            border: 1px solid #e6e6e6;
            border-radius: 3px;
            line-height: 33px;
            text-align: center;
            box-sizing: border-box;
        }
        .city_body .city_sort div{
            margin-top: 20px;
        }
        .city_body .city_sort h2{
            padding-left: 15px;
            line-height: 30px;
            font-size: 14px;
            background: #f0f0f0;
            font-weight: normal;
        }
        .city_body .city_sort ul{
            padding-left: 10px;
            margin-top: 10px;
        }
        .city_body .city_sort ul li{
```

```
        line-height: 30px;
    }
    .city_body .city_index{
        width: 20px;
        display: flex;
        flex-direction: column;
        justify-content: center;
        text-align: center;
        border-left:1px solid #e6e6e6;
    }
</style>
```

城市组件在谷歌浏览器中的运行效果如图 13-20 所示。

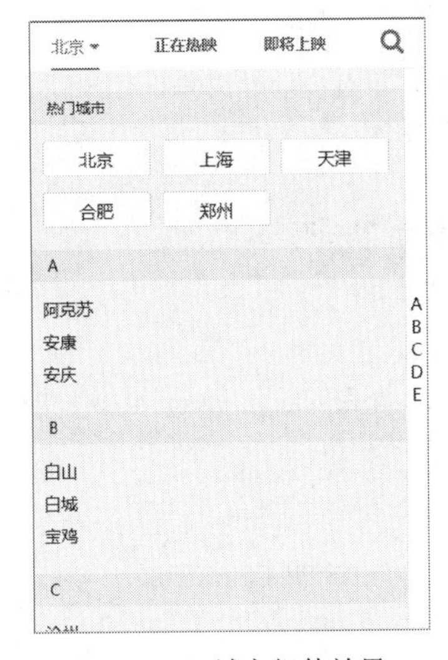

图 13-20 城市组件效果

2. 正在热映组件（NowPlaying）

正在热映组件是由一个列表设计完成，代码如下：

```
<template>
    <div class="movie_body">
        <ul>
            <li>
                <div class="pic_show"><img src="../../../public/images/001.png"
                   alt=""></div>
                <div class="info_list">
                    <h2>机械师2：复活</h2>
                    <p>观众评<span class="grade"> 8.9</span></p>
                    <p>主演： 杰森·斯坦森 杰西卡·阿尔芭 汤米·李·琼斯 杨紫琼 山姆·哈兹
                       尔丁</p>
                    <p>今天50家影院放映800场</p>
                </div>
                <div class="btn_mall">
                    购票
                </div>
            </li>
            <li>
                <div class="pic_show"><img src="../../../public/images/002.png"
                   alt=""></div>
```

```
            <div class="info_list">
                <h2>敢死队</h2>
                <p>观众评<span class="grade"> 8.7</span></p>
                <p>主演： 西尔维斯特·史泰龙，杰森·斯坦森，梅尔·吉布森</p>
                <p>今天50家影院放映750场</p>
            </div>
            <div class="btn_mall">
                购票
            </div>
        </li>
        <li>
            <div class="pic_show"><img src="../../../public/images/003.png"
              alt=""></div>
            <div class="info_list">
                <h2>最后的巫师猎人</h2>
                <p>观众评<span class="grade"> 8.4</span></p>
                <p>主演： 范·迪塞尔，萝斯·莱斯利，伊利亚·伍德，迈克尔·凯恩，丽纳·欧
                  文</p>
                <p>今天50家影院放映600场</p>
            </div>
            <div class="btn_mall">
                购票
            </div>
        </li>
        <li>
            <div class="pic_show"><img src="../../../public/images/004.png"
              alt=""></div>
            <div class="info_list">
                <h2>饥饿游戏3</h2>
                <p>观众评<span class="grade"> 7.6</span></p>
                <p>主演： 詹妮弗·劳伦斯，乔什·哈切森，利亚姆·海姆斯沃斯</p>
                <p>今天50家影院放映550场</p>
            </div>
            <div class="btn_mall">
                购票
            </div>
        </li>
        <li>
            <div class="pic_show"><img src="../../../public/images/005.png"
              alt=""></div>
            <div class="info_list">
                <h2>钢铁骑士</h2>
                <p>观众评<span class="grade"> 7.3</span></p>
                <p>主演： 本·温切尔，乔什·布雷纳，玛丽亚·贝罗， 迈克·道尔， 安迪·
                  加西亚</p>
                <p>今天50家影院放映500场</p>
            </div>
            <div class="btn_mall">
                购票
            </div>
        </li>
        <li>
            <div class="pic_show"><img src="../../../public/images/006.png"
              alt=""></div>
            <div class="info_list">
                <h2>奔跑者
                </h2>
                <p>观众评<span class="grade"> 6.6</span></p>
                <p>主演： 尼古拉斯·凯奇，康妮·尼尔森，莎拉·保罗森，彼得·方达</p>
                <p>今天50家影院放映500场</p>
```

```
                </div>
                <div class="btn_mall">
                    购票
                </div>
            </li>
        </ul>
    </div>
</template>
<script>
    export default {
        name: "NowPlaying"
    }
</script>
<style scoped>
    #content .movie_body{
        flex: 1;overflow: auto;
    }
    .movie_body ul{
        margin: 0 12px;
        overflow: hidden;
    }
    .movie_body ul li{margin-top: 12px;display: flex;align-items:
      center;border-bottom: 1px solid #e6e6e6;padding-bottom: 10px;}
    .movie_body .pic_show{width: 64px;height: 90px;}
    .movie_body .pic_show img{width: 100%;}
    .movie_body .info_list{margin-left:10px;flex: 1;position: relative; }
    .movie_body .info_list h2{
        font-size: 17px; line-height: 24px;
        width: 150px;overflow: hidden;
        white-space: nowrap;
        text-overflow:ellipsis ;
    }
    .movie_body .info_list p{
        font-size:13px;
        color: #666;
        line-height: 22px;
        width: 200px;
        overflow: hidden;
        white-space: nowrap;
        text-overflow:ellipsis ;
    }
    .movie_body .info_list .grade{
        font-weight: 700;
        color: #faaf00;
        font-size: 15px;
    }
    .movie_body .info_list img{
        width: 50px;
        position: absolute;
        right: 10px;
        top: 5px;
    }
    .movie_body .btn_mall, .movie_body .btn_pre{
        width: 47px;
        height: 27px;
        line-height: 28px;
        text-align: center;
        background-color: #f03d37;
        color: #fff;
        border-radius: 4px;
```

```
        font-size: 12px;
        cursor: pointer;
    }
    .movie_body .btn_pre{
        background-color: #3c9fe6;
    }
</style>
```

正在热映组件在谷歌浏览器中的运行效果如图 13-21 所示。

图 13-21　正在热映组件效果

3. 即将上映组件（ComingSoon）

即将上映组件和正在热映组件类似，由一个列表组成，代码如下：

```
<template>
    <div class="movie_body">
        <ul>
            <li>
                <div class="pic_show"><img src="../../../public/images/007.png"
                  alt=""></div>
                <div class="info_list">
                    <h2>佐罗和麦克斯</h2>
                    <p><span class="person">46465</span>人想看</p>
                    <p>主演：格兰特·鲍尔，艾米·斯马特，博伊德·肯斯特纳</p>
                    <p>未来30天内上映</p>
                </div>
                <div class="btn_pre">
                    预售
                </div>
            </li>
            <li>
                <div class="pic_show"><img src="../../../public/images/008.png"
                  alt=""></div>
                <div class="info_list">
                    <h2>废材特工</h2>
                    <p><span class="person">64645</span>人想看</p>
                    <p>主演：  杰西·艾森伯格，克里斯汀·斯图尔特，约翰·雷吉扎莫</p>
                    <p>未来30天内上映</p>
```

```
            </div>
            <div class="btn_pre">
                预售
            </div>
        </li>
        <li>
            <div class="pic_show"><img src="../../../public/images/009.png"
              alt=""></div>
            <div class="info_list">
                <h2>凤凰城遗忘录</h2>
                <p><span class="person">42465</span>人想看</p>
                <p>主演: Clint Jordan</p>
                <p>未来30天内上映</p>
            </div>
            <div class="btn_pre">
                预售
            </div>
        </li>
        <li>
            <div class="pic_show"><img src="../../../public/images/010.png"
              alt=""></div>
            <div class="info_list">
                <h2>新灰姑娘</h2>
                <p><span class="person">46465</span>人想看</p>
                <p>主演: Cassandra Morris, Kristen Day</p>
                <p>未来30天内上映</p>
            </div>
            <div class="btn_pre">
                预售
            </div>
        </li>
        <li>
            <div class="pic_show"><img src="../../../public/images/011.png"
              alt=""></div>
            <div class="info_list">
                <h2>鲨卷风4: 四度觉醒</h2>
                <p><span class="person">38465</span>人想看</p>
                <p>主演: 塔拉·雷德, Ian Ziering, Masiela Lusha</p>
                <p>未来30天内上映</p>
            </div>
            <div class="btn_pre">
                预售
            </div>
        </li>
        <li>
            <div class="pic_show"><img src="../../../public/images/012.png"
              alt=""></div>
            <div class="info_list">
                <h2>全境警戒</h2>
                <p><span class="person">46465</span>人想看</p>
                <p>主演: 戴夫·巴蒂斯塔, 布兰特妮·斯诺, Angelic Zambrana</p>
                <p>未来30天内上映</p>
            </div>
            <div class="btn_pre">
                预售
            </div>
        </li>
    </ul>
  </div>
</template>
```

```
<script>
    export default {
        name: "ComingSoon"
    }
</script>
<style scoped>
    #content .movie_body{
        flex: 1;overflow: auto;
    }
    .movie_body ul{
        margin: 0 12px;
        overflow: hidden;
    }
      .movie_body ul li{margin-top: 12px;display: flex;align-items:
         center;border-bottom: 1px solid #e6e6e6;padding-bottom: 10px;}
    .movie_body .pic_show{width: 64px;height: 90px;}
    .movie_body .pic_show img{width: 100%;}
    .movie_body .info_list{margin-left:10px;flex: 1;position: relative; }
    .movie_body .info_list h2{
        font-size: 17px; line-height: 24px;
        width: 150px;overflow: hidden;
        white-space: nowrap;
        text-overflow:ellipsis ;
    }
    .movie_body .info_list p{
        font-size:13px;
        color: #666;
        line-height: 22px;
        width: 200px;
        overflow: hidden;
        white-space: nowrap;
        text-overflow:ellipsis ;
    }
    .movie_body .info_list .grade{
        font-weight: 700;
        color: #faaf00;
        font-size: 15px;
    }
    .movie_body .info_list img{
        width: 50px;
        position: absolute;
        right: 10px;
        top: 5px;
    }
    .movie_body .btn_mall, .movie_body .btn_pre{
        width: 47px;
        height: 27px;
        line-height: 28px;
        text-align: center;
        background-color: #f03d37;
        color: #fff;
        border-radius: 4px;
        font-size: 12px;
        cursor: pointer;
    }
    .movie_body .btn_pre{
        background-color: #3c9fe6;
    }
</style>
```

即将上映组件在谷歌浏览器中的运行效果如图 13-22 所示。

图 13-22　即将上映组件效果

4. 搜索组件（Search）

搜索组件代码如下：

```
<template>
    <div class="search_body">
        <div class="search_input">
            <div class="search_input_wrapper">
                <i class="fa fa-search"></i>
                <input type="text">
            </div>
        </div>
        <div class="search_result">
            <h3>电影/电视剧/综艺</h3>
            <ul>
                <li>
                    <div class="img"><img src="../../../public/images/001.png"
                       alt=""></div>
                    <div class="info">
                        <p><span>机械师2 </span><span>8.9</span></p>
                        <p>剧情，喜剧，犯罪</p>
                        <p>2020-6-30</p>
                    </div>
                </li>
            </ul>
        </div>
    </div>
</template>
<script>
    export default {
        name: "Search"
    }
</script>
<style scoped>
    #content .search_body{
        flex: 1;
```

```
    overflow: auto;
}
.search_body .search_input{
    padding: 8px 10px;
    background-color: #f5f5f5;
    border-bottom: 1px solid #e5e5e5;
}
.search_body .search_input_wrapper{
    padding: 0 10px;
    border: 1px solid #e6e6e6;
    border-radius: 5px;
    background-color: #fff;
    display: flex;
}
.search_body .search_input_wrapper i{
    font-size: 16px;
    padding: 4px 0;
}
.search_body .search_input_wrapper input{
    border: none;
    font-size: 13px;
    color: #333;
    padding: 4px 0;
    outline: none;
}
.search_body .search_result h3{
    font-size: 15px;
    color: #999;
    padding: 9px 15px;
    border-bottom: 1px solid #e6e6e6;
}
.search_body .search_result li{
    border-bottom: 1px #c9c9c9 dashed;
    padding: 10px 15px;
    box-sizing: border-box;
    display: flex;
}
.search_body .search_result .img{
    width: 60px;
    float: left;
}
.search_body .search_result .img img{
    width: 100%;
}
.search_body .search_result .info{
    float: left;
    margin-left: 15px;
    flex: 1;
}
.search_body .search_result .info p{
    height: 22px;
    display: flex;
    line-height: 22px;
    font-size: 12px;
}
.search_body .search_result .info p:nth-of-type(1) span:nth-of-type(1){
    font-size: 18px;
    flex: 1;
}
```

```
    .search_body .search_result .info p:nth-of-type(1) span:nth-of-type(2){
        font-size: 16px;
        color: #fc7103;
    }
</style>
```

搜索组件在谷歌浏览器中的运行效果如图 13-23 所示。

图 13-23　搜索组件效果

13.3.3　设计影院页面组件

影院页面只有一个组件——影院列表组件（CiList），也是由一个列表组设计完成。代码如下：

```
<template>
    <div class="cinema_body">
        <ul>
            <li>
                <div>
                    <span>大地影院延庆金锣湾店</span>
                    <span class="q"><span class="price"> 38.5</span> 元起</span>
                </div>
                <div class="address">
                    <span>延庆区北街39号H座首层</span>
                    <span> >100km </span>
                </div>
                <div class="card">
                    <div>小吃</div>
                    <div>折扣卡</div>
                </div>
            </li>
            <li>
                <div>
                    <span>燕山影剧院</span>
                    <span class="q"><span class="price"> 37.5</span> 元起</span>
                </div>
```

```
                <div class="address">
                    <span>房山区燕山岗南路3号</span>
                    <span> >120km</span>
                </div>
                <div class="card">
                    <div>小吃</div>
                    <div>折扣卡</div>
                </div>
            </li>
            <li>
                <div>
                    <span>万达影城昌平保利光魔店</span>
                   <span class="q"><span class="price"> 37.9</span> 元起</span>
                </div>
                <div class="address">
                    <span>昌平区鼓楼南街佳莲时代广场四层</span>
                    <span> >80km </span>
                </div>
                <div class="card">
                    <div>小吃</div>
                    <div>折扣卡</div>
                </div>
            </li>
            <li>
                <div>
                    <span>门头沟影剧院</span>
                   <span class="q"><span class="price"> 30.9</span> 元起</span>
                </div>
                <div class="address">
                    <span>门头沟区新桥大街12号</span>
                    <span>  >110km </span>
                </div>
                <div class="card">
                    <div>小吃</div>
                    <div>折扣卡</div>
                </div>
            </li>
        </ul>
    </div>
</template>
<script>
    export default {
        name: "CiList"
    }
</script>
<style scoped>
    #content .cinema_body{
        flex: 1;
        overflow: auto;
    }
    .cinema_body ul{
        padding: 20px;
    }
    .cinema_body li{
        border-bottom: 1px solid #e6e6e6;
        margin-bottom: 20px;
    }
    .cinema_body div{
        margin-bottom: 10px;
```

```
    }
    .cinema_body .q{
        font-size: 11px;
        color: #f03d37;
    }
    .cinema_body .price{
        font-size: 18px;
    }
    .cinema_body .address{
        font-size: 13px;
        color:#666;
    }
    .cinema_body .address span:nth-of-type(2){
        float: right;
    }
    .cinema_body .card{
        display: flex;
    }
    .cinema_body .card div{
        padding: 0 3px;
        height: 15px;
        line-height: 15px;
        border-radius:2px;
        color: #f90;
        border:1px solid #f90;
    }
    .cinema_body .card div.or{
        color: #f90;
        border: 1px solid #f90;
    }
    .cinema_body .card div.bl{
        color: #589daf;
        border: 1px solid #589daf;
    }
</style>
```

影院页面组件在谷歌浏览器中的运行效果如图 13-24 所示。

图 13-24　影院页面组件效果

13.3.4 设计我的页面组件

我的页面只有一个登录/注册组件，这里只是一个简单的登录/注册表单，并没有实现前后端的交互。代码如下：

```
<template>
    <div class="login_body">
        <div>
            <input class="login_text" type="text" placeholder="账号/手机号/邮箱">
        </div>
        <div>
            <input class="login_text" type="password" placeholder="请输入您的密码">
        </div>
        <div class="login_btn">
            <input type="submit" value="登录">
        </div>
        <div class="login_link">
            <a href="#">立即注册</a>
            <a href="#">找回密码</a>
        </div>
    </div>
</template>

<script>
    export default {
        name: "Login"
    }
</script>

<style scoped>
    #content .login_body{
        width: 100%;
    }
    .login_body .login_text{
        width: 100%;
        height: 40px;
        border: none;
        border-bottom: 1px #ccc solid;
        margin:0 5px;
        outline: none;
    }
    .login_body .login_btn{
        height: 50px;
        margin: 10px;
    }
    .login_body .login_btn input{
        display: block;
        width: 100%;
        height: 100%;
        background: #e54847;
        border-radius: 3px;
        border: none;
        color: white;
    }
    .login_body .login_link{
        display: flex;
        justify-content: space-between;
    }
    .login_body .login_link a{
```

```
        text-decoration: none;
        margin: 0 5px;
        font-size: 12px;
        color:#e54847;
    }
</style>
```

我的页面组件在谷歌浏览器中的运行效果如图 13-25 所示。

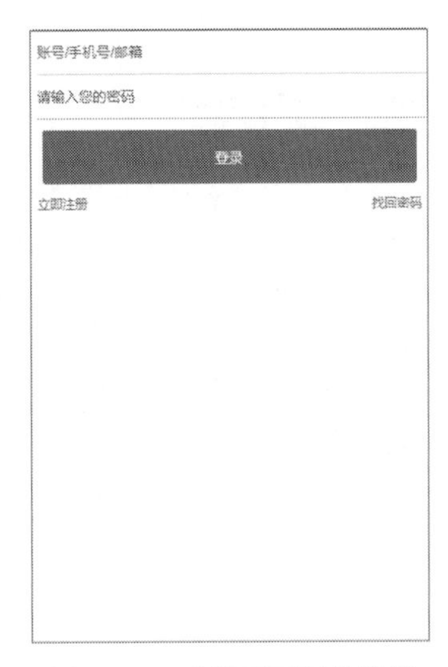

图 13-25　我的页面组件效果

13.4 设计项目页面组件及路由配置

前面已经介绍了组成电影页面、影院页面和我的页面所包含的所有组件，接下来就是组合它们。

13.4.1　电影页面组件及路由

电影页面（Movie）顶部有 4 个导航元素，对应着前面定义的城市、正在热映、即将上映和搜索等组件，使用 <router-link> 标签进行导航切换。

```
<template>
    <div id="main">
        <!-- 头部组件-->
        <Header title="神影电影"></Header>
        <div id="content">
            <div class="movie_menu">
                <router-link tag="div" to="/movie/city" class="city_name">
                    <span>北京 </span><i class="fa fa-caret-down"></i>
                </router-link>
                <div class="hot_swtich">
```

```
                <router-link tag="div" to="/movie/nowPlaying" class="hot_
                  item active">正在热映</router-link>
                <router-link tag="div" to="/movie/comingSoon" class="hot_
                  item">即将上映</router-link>
            </div>
            <router-link tag="div" to="/movie/search" class="search_entry">
                <i class="fa fa-search"></i>
            </router-link>
        </div>
        <!--二级路由渲染-->
        <keep-alive>
            <router-view></router-view>
        </keep-alive>

    </div>
    <!-- 尾部组件-->
    <TabBar></TabBar>
  </div>
</template>
<script>
    //引入组件
    import Header from '../../components/Header';
    import TabBar from '../../components/TabBar';
    export default {
        name:'Movie',
        components:{
            Header,
            TabBar
        }
    }
</script>
<style scoped>
    #content .movie_menu{
        width: 100%;
        height: 45px;
        border-bottom: 1px solid #e6e6e6;
        display: flex;
        justify-content: space-between;
    }
    .movie_menu .city_name{
        margin-left: 20px;
        height: 100%;
        line-height: 45px;
    }
    .movie_menu .city_name.router-link-active{
        color: #ef4238;
        border-bottom: 2px solid #ef4238;
        box-sizing: border-box;
    }
    .movie_menu .hot_swtich{
        display: flex;
        height: 100%;
        line-height: 45px;
    }
    .movie_menu .hot_item{
        font-size: 15px;
        color: #666;
        width: 80px;
        text-align: center;
        margin: 0 12px;
```

```
        font-weight: 700;
    }
    .movie_menu .hot_item.router-link-active{
        color: #ef4238;
        border-bottom:2px solid #ef4238;
    }
    .movie_menu .search_entry{
        margin-right: 20px;
        height: 100%;
        line-height: 45px;
    }
    .movie_menu .search_entry.router-link-active{
        color: #ef4238;
        border-bottom:2px solid #ef4238;
        box-sizing: border-box;
    }
    .movie_menu .search_entry i{
        font-size: 24px;
        color: red;
    }
</style>
```

配置路由，代码如下：

```
//movie路由
export default {
    path:'/movie',
    //按需载入的方式
    component:()=>import('../../views/Movie'),
    //二级路由，使用children进行配置
    children:[
        {
            path:'city',
            component:()=>import('../../components/City')
        },
        {
            path:'nowPlaying',
            component:()=>import('../../components/NowPlaying')
        },
        {
            path:'comingSoon',
            component:()=>import('../../components/ComingSoon')
        },
        {
            path:'search',
            component:()=>import('../../components/Search')
        },
        //重定向：当路径为/movie时，重定向到/movie/nowPlaying路径
        {
            path:'/movie',
            redirect:'/movie/nowPlaying'
        }
    ]
}
```

依次切换城市组件、正在热映组件、即将上映组件和搜索组件，效果如图 13-26 ～图 13-29 所示。

图 13-26　城市组件

图 13-27　正在热映组件

图 13-28　即将上映组件

图 13-29　搜索组件

13.4.2　影院页面组件及路由

影院页面组件中直接引入头部组件、底部导航组件和影院列表组件，代码如下：

```
<template>
    <div id="main">
        <Header title="神影影院"></Header>
        <div id="content">
            <div class="cinema_menu">
                <div class="city_switch">
                    全城 <i class="fa fa-caret-down"></i>
                </div>
                <div class="city_switch">
                    品牌 <i class="fa fa-caret-down"></i>
                </div>
                <div class="city_switch">
                    特色 <i class="fa fa-caret-down"></i>
                </div>
            </div>
            <CiList></CiList>
        </div>
        <TabBar></TabBar>
    </div>
</template>
<script>
    import Header from '../../components/Header';
    import TabBar from '../../components/TabBar';
    import CiList from '../../components/CiList';
    export default {
        name:'Cinema',
        components:{
            Header,
            TabBar,
            CiList
        }
    }
</script>
<style scoped>
#content .cinema_menu{
    width: 100%;
    height: 45px;
    border-bottom: 1px solid #e6e6e6;
    display: flex;
    justify-content: space-around;
    align-items: center;
    background: white;
}
</style>
```

路由配置如下：

```
//Cinema路由
export default {
    path:'/cinema',
    component:()=>import('../../views/Cinema')
}
```

影院页面组件在谷歌浏览器中的运行效果如图 13-30 所示。

图 13-30　影院页面组件效果

13.4.3　我的页面组件及路由

在影院页面组件中直接引入头部组件、底部导航组件和影院列表组件，代码如下：

```
<template>
    <div id="main">
        <Header title="我的神影"></Header>
        <div id="content">
            <Login></Login>
        </div>
        <TabBar></TabBar>
    </div>
</template>
<script>
    import Header from '../../components/Header';
    import TabBar from '../../components/TabBar';
    import Login from '../../components/Login';
    export default {
        name:'Mine',
        components:{
            Header,
            TabBar,
            Login
        }
    }
</script>
<style scoped></style>
```

路由配置如下：

```
//mine路由
export default {
    path:'/mine',
    component:()=>import('../../views/Mine')
}
```

我的页面组件在谷歌浏览器中的运行效果如图 13-31 所示。

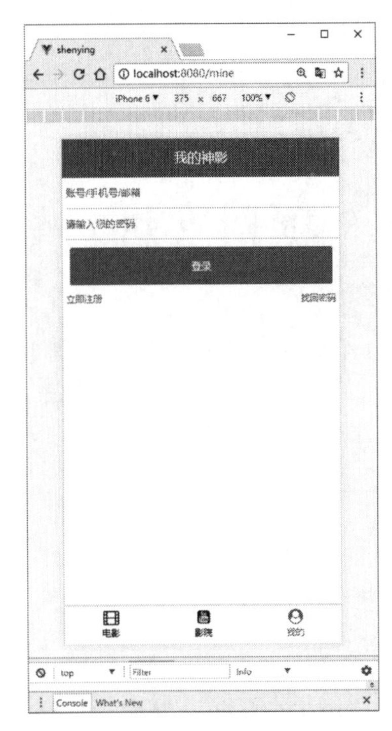

图 13-31　我的页面组件效果

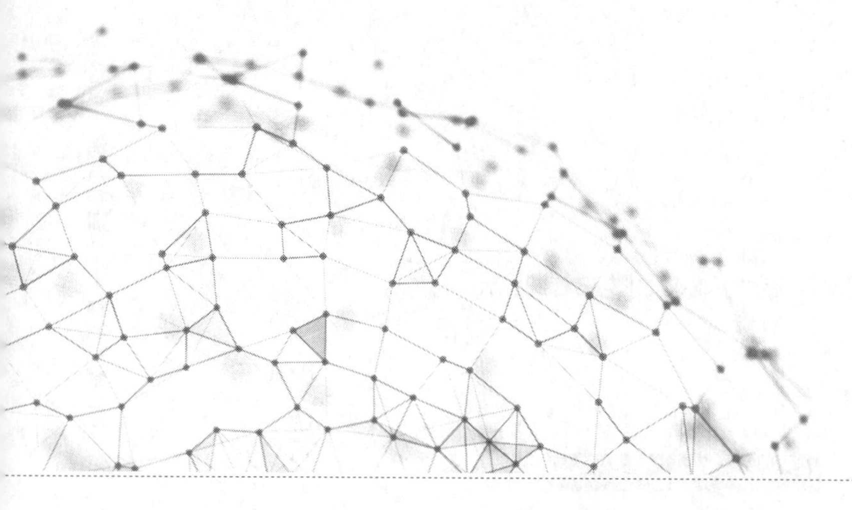

第14章

项目实训2——音乐之家App

本项目是一个移动端的音乐应用程序，使用 Vue 框架设计开发。项目是使用 Vue-CLI 进行搭建，在项目中还使用了 Vue 的相关插件，例如：路由管理器（vue-router）、状态管理（vuex）。

14.1 项目概述

项目是一个基于 Vue 2.0 版本的音乐 App，包含搜索和播放音乐等功能。

14.1.1 开发环境

需要安装 Node.js 和 NPM，一般情况下 Node.js 中已经集成了 NPM。然后安装 Vue 脚手架（Vue-CLI）以及创建项目，具体的安装步骤请参考“脚手架”这一章。

对于项目的调试，是在谷歌浏览器的控制台进行模拟。打开浏览器后，按键盘上的 F12 键，然后单击“切换设备工具栏”，进入移动端的调试界面，可以选择相应的设备进行调试，如图 14-1 所示。

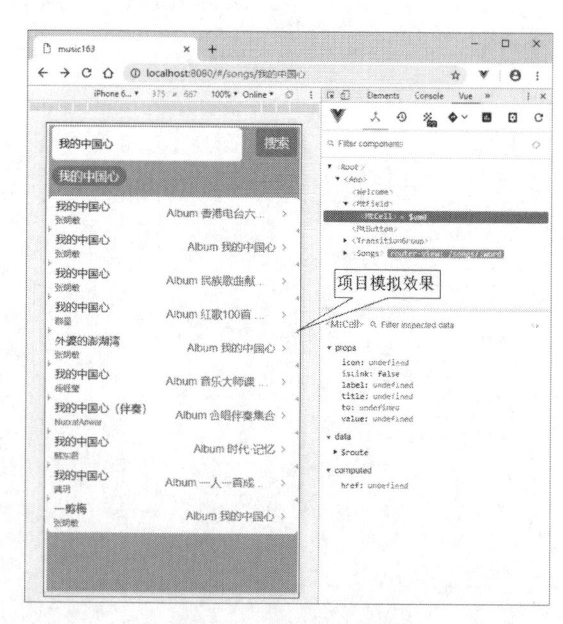

图 14-1　项目调试界面

14.1.2 技术概括

项目主要使用的技术，说明如下。

- Vue 2.0：是一套构建用户界面、以视图为核心、数据为驱动的组件化框架。只要把数据传入已经编译好的模板中，便能渲染出想要的视图。
- Vue-CLI：Vue-CLI是Vue官方支持的一个脚手架，会随本版本进行迭代更新。它有一套成熟的Vue项目架构设计，能够快速初始化一个Vue项目。还提供了一套本地的node测试服务器，使用Vue-CLI自己提供的命令，就可以启动服务器。可以集成打包上线方案。还有一些优点，包括模块化、转译、预处理、热加载、静态检测和自动化测试等。
- Vue-router：是Vue官方的路由管理器。它和Vue的核心深度集成，让构建单页面应用变得易如反掌。
- Vuex：是一个专为Vue应用程序开发的状态管理模式。它采用集中式存储管理应用的所有组件的状态，并以相应的规则保证状态以一种可预测的方式发生变化。
- mint-ui：基于Vue的移动端组件库。
- vue-resource：vue-resource是Vue的一款插件，它可以通过XMLHttpRequest或JSONP发起请求并处理响应。

14.1.3 项目结构

项目结构如图14-2所示，其中src文件夹是项目的源码目录，如图14-3所示。

图14-2 项目结构

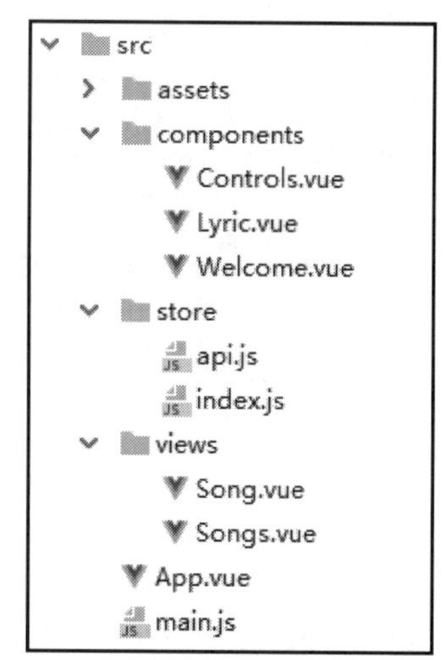

图14-3 src文件夹结构

项目结构中，主要文件说明如下。

- build：构建服务和webpack配置，转发聊天机器人及ajax获取用户数据相关内容。
- config：项目不同环境的配置。

- index.html：项目入口页面。
- package.json：项目配置文件。

src 文件夹目录说明如下。

- assets：字体图标及 CSS 样式文件夹。
- components：项目组件文件夹。
- store：存放状态管理 Vuex 的相关文件夹。
- views：项目组件文件夹（页面级）。
- App.vue：模板入口文件。
- main.js：程序入口文件，路由的配置也在其中。

14.2 入口文件

入口文件有 index.html、main.html 和 App.vue 三个文件。下面分别介绍。

14.2.1 项目入口页面（index. html）

index.html 是项目默认的主渲染页面文件，主要是一些引用文件。具体代码如下：

```
<!DOCTYPE html>
<html lang=zh-cmn-Hans>
  <head>
    <meta charset="utf-8">
    <title>vue-music</title>
    <meta name="screen-orientation" content="portrait">
    <meta name="full-screen" content="yes">
    <meta name="browsermode" content="application">
    <meta name="x5-orientation" content="portrait">
    <meta name="x5-fullscreen" content="true">
    <meta name="x5-page-mode" content="app">
      <meta name="viewport" content="width=device-width,initial-scale=1.0,
        minimum-scale=1.0, maximum-scale=1.0, user-scalable=no,minimal-ui">
    <style type="text/css">
      body{
        margin:0;
        padding: 0;
        height:100%;
      }
    </style>
      <link rel="stylesheet" href="https://cdn.bootcss.com/font-
        awesome/5.10.0-11/css/all.css" >
  </head>
  <body>
    <div id="app"></div>
  </body>
</html>
```

14.2.2 程序入口文件（main.js）

main.js是程序入口文件，加载各种公共组件以及初始化Vue实例，本项目的路由也在该文件中进行配置。具体代码如下：

```
import Vue from 'vue';
import VueRouter from 'vue-router';
import App from './App';
import MintUI from 'mint-ui';
import 'mint-ui/lib/style.css';
import vueTap from 'v-tap';
import store from './store';
Vue.use(vueTap);
Vue.use(VueRouter);
Vue.use(MintUI);
//配置路由
const router = new VueRouter({
  //路由模式
  mode: 'hash',
  base: __dirname,
  routes: [
    { path: '/' },
    { path: '/songs/:word', component: require('./views/Songs') },
    { path: '/song/:id', component:require('./views/Song') },
    { path: '*', redirect: '/' }
  ]
})
var vm=new Vue({
  el: '#app',
  template: '<App/>',
  components: { App },
  router,
  store
})
```

14.2.3 组件入口文件（App.vue）

App.vue是项目的根组件，所有的页面都是在App.vue下面切换的，可以理解为所有组件都是App.vue的子组件。除了引用相应的组件外，项目的搜索功能也是在根组件实现。

```
<template>
  <div id="app" class="container-flex box-ver">
    <!-- 引入welcome组件-->
    <welcome @afterLeave="toSongs"></welcome>
    <form id="form-app" @submit.prevent="handleSearch()">
      <div class="container-flex">
        <div class="f1">
          <mt-field v-model="value" placeholder="输入歌曲名或歌手名"></mt-field>
        </div>
        <div style="margin-left:20px;" v-tap="{methods:handleSearch}">
            <mt-button type="primary">搜索</mt-button>
        </div>
      </div>
```

```
      <div class="my-badges">
        <transition-group name="list">
          <mt-badge type="primary" size="large" v-for="word of searchWordArr"
            :key="word">
            <span @click="handleWord(word)">{{word}}</span>
          </mt-badge>
        </transition-group>
      </div>
    </form>
    <router-view class="view"></router-view>
  </div>
</template>
<script>
import Welcome from './components/Welcome'
import { Toast } from 'mint-ui';
export default {
  name: 'app',
  data:function(){
    return {
      value:'',
      searchWordArr:this.$store.state.searchWordArr
    }
  },
  components: {
    Welcome
  },
  methods:{
    handleSearch(){
      console.log("click:"+this.value);
      if (this.value=='') {
        Toast('搜索关键字不可为空');
        return;
      }
      if(!this.$store.state.showWelcome){
        this.toSongs();
      }else{
        this.$store.commit("hideWelcome");
      }
    },
    handleWord(word){
      this.value=word;
      this.handleSearch();
    },
    toSongs(){
      if (this.value=='') {
        return;
      }
      console.log("run toSongs")
      this.$router.push({ path: '/songs/'+this.value});
      this.$store.commit("addSearchWord",{value:this.value});
    }
  }
}
</script>
<style>
#app {
  font-family: 'Avenir', Helvetica, Arial, sans-serif;
  -webkit-font-smoothing: antialiased;
  -moz-osx-font-smoothing: grayscale;
```

```
  color: #2c3e50;
  background: #6ecc9e;
height: 100vh;
}
a{
  text-decoration:none;
}
.mint-cell{
  border-radius:0.4em;
}
.container-flex{
  display: flex;
  /* position:relative; */
  margin: 2px 4px;
}
.f1{
  position:relative;
  flex: 1;
}
.box-ver{
  flex-direction:column;
}
.my-badges{
  margin:0.4em 0 0.8em 0;
  overflow: hidden;
}
.mint-badge{
  margin:0.1em 0.2em;
}
.mint-cell-title{
  min-width: 6em;
}
.slide-fade-enter-active {
  transition: all .5s ease;
}
.slide-fade-leave-active {
  transition: all .3s cubic-bezier(1.0, 0.5, 0.8, 1.0);
}
.slide-fade-enter, .slide-fade-leave-active {
  padding-left: 10px;
  opacity: 0;
}
.list-enter-active, .list-leave-active {
  transition: all 1s;
}
.list-enter, .list-leave-active {
  opacity: 0;
  transform: translateY(30px);
}
@media (min-width: 1024px) {
    #app {
        width:480px;
        min-height:800px;
        padding:10px;
        border:1px solid #030303;
        border-radius: 0.3em;
        margin:0 auto;
    }
}
</style>
```

14.3 状态管理

Vuex 是一个专为 Vue 应用程序开发的状态管理插件。它就像一个“仓库”，管理项目组件共享的数据。

14.3.1 store/api.js

在 api.js 文件中，使用 vue-resource 请求歌曲的相关数据，使用 querystring 转换请求的字符串代码。具体实现代码如下：

```
import Vue from 'vue';
import VueResource from 'vue-resource';
var querystring = require('querystring');
Vue.use(VueResource)
Vue.http.options.emulateJSON = true;
export function fetchSongList (options) {
  var otherParams ={
        'csrf_token':"",
        'type':1,
        'offset':0,
        'limit':10,
        'total':true
    };
    var postData = Object.assign(options, otherParams);
    console.log(postData);
    //vue-resource 当需要发送大量的参数到服务器的时候，需要使用post请求
    var p=Vue.http.post("/api163/search/get/web",postData);
    p.then(resp => {
        console.log(resp.data);
    }, resp => {
        console.log("request error");
    });
  return p;
}
export function fetchSong (music_id) {
  var p=Vue.http.get("/api163/song/detail?id="+music_id+"&ids="+'%5B'+music_
    id+'%5D');
      console.log('url',"http://music.163.com/api/song/detail?id="+music_
        id+"&ids="+'%5B'+music_id+'%5D')
    p.then(resp=>{
      console.log(resp.data);
    },resp=>{
      console.log("request error");
    });
    return p;
}
export function fetchLyric (music_id) {
  var p=Vue.http.get("/api163/song/lyric?os=pc&id="+music_id+'&lv=-1&kv=-
    1&tv=-1');
    p.then(resp=>{
      console.log(resp.data);
    },resp=>{
      console.log("request error");
```

```
    });
    return p;
}
```

14.3.2 store/index.js

状态管理有 5 个核心，分别是 state、getter、mutation、action 以及 module，在本项目中全在 index.js 中进行定义。具体的代码如下：

```
import Vue from 'vue'
import Vuex from 'vuex'
import { fetchSongList, fetchSong, fetchLyric } from './api'
import _ from 'lodash'
Vue.use(Vuex);
const store = new Vuex.Store({
    state: {
        showWelcome: true,
        searchWordArr: localStorage.searchWords ? JSON.parse(localStorage.
          searchWords) : [],
        songs: [],
        song: null,
        lyricArr: [],     //歌词
        lrcTimeArr: [],   //时间
        lrcMarginTop:0,
        playing:false
    },
    getters: {},
    actions: {
        FETCH_SONG_LIST(context, options) {
            let p = fetchSongList(options);
            p.then(resp => {
               context.commit("loadSongList", { songs: resp.data.result.songs });
            });
            return p;
        },
        FETCH_SONG(context, music_id) {
            let p = fetchSong(music_id);
            p.then(resp => {
                context.commit("loadSong", { song: resp.data.songs[0] });
            });
            return p;
        },
        FETCH_LYRIC(context, music_id) {
            let p = fetchLyric(music_id);
            p.then(resp => {
                context.commit("loadLyric", { lyric: resp.data.lrc.lyric });
            });
            return p;
        },
    },
    mutations: {
        hideWelcome(state) {
            state.showWelcome = false;
        },
        addSearchWord(state, payload) {
            if (state.searchWordArr.indexOf(payload.value) < 0) {
                state.searchWordArr.unshift(payload.value);
```

```
            }
            if (state.searchWordArr.length >5) {
                state.searchWordArr = state.searchWordArr.slice(0, 5);
            }
            localStorage.searchWords = JSON.stringify(state.searchWordArr);
        },
        loadSongList(state, payload) {
            state.songs = payload.songs;
        },
        loadSong(state, payload) {
            state.song = payload.song;
        },
        loadLyric(state, payload) {
            let arr = payload.lyric.split('\n');
            state.lyricArr = convertLrcArr(arr);
            state.lrcTimeArr=_.map(state.lyricArr,'time');
        },
        changePlaying(state,isPlaying){
            state.playing=isPlaying;
        },
        changeLrcMarginTop(state,lrcMarginTop){
            state.lrcMarginTop=lrcMarginTop;
        }
    }
});
function convertLrcArr(arr) {
    let lrcArr = [];
    let duration = 0;
    console.log(arr.length)
    for (let i = 0; i < arr.length - 1; i++) {
        let item = arr[i];
        let lrcObj = {};
        let timeStr = item.match("\\[(.+?)\\]")[1];
        //declude not time
        if (/[^0-9\.\:]/.test(timeStr)) {
            continue;
        }
        let timeArr = timeStr.split(":");
        let time = parseInt(timeArr[0]) * 60 + parseFloat(timeArr[1]);
        lrcObj.selected = false;
        lrcObj.show = true;
        lrcObj.time = time;
        lrcObj.lrc = item.replace(new RegExp(/(\.\d{2,3})/g), '');
        lrcArr.push(lrcObj);
    }
    return lrcArr;
}
export default store
```

14.4 项目组件设计

项目的组件在 components 文件夹和 view 文件夹中定义。

14.4.1 欢迎组件

欢迎组件（Welcome.vue）中定义了 App 的名称，以及欢迎语句。在谷歌浏览器控制台模拟，欢迎组件效果如图 14-4 所示。

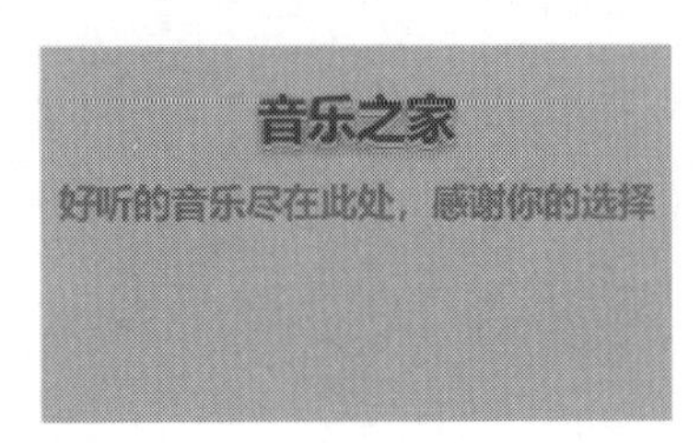

图 14-4　欢迎组件效果

具体的代码如下：

```
<template>
  <transition name="slide-up" v-on:after-leave="afterLeave">
  <div v-show="show" class="welcome neonText">
    <h1>{{ msg }}</h1>
    <h2>好听的音乐尽在此处，感谢你的选择</h2>
  </div>
  </transition>
</template>
<script>
export default {
  name: 'welcome',
  computed:{
    show(){
      return this.$store.state.showWelcome;
    }
  },
  data () {
    return {
      msg: '音乐之家'
    }
  },
  methods:{
    afterLeave(){
      this.$emit("afterLeave");
    }
  }
}
</script>
<!-- Add "scoped" attribute to limit CSS to this component only -->
<style scoped>
.welcome{
  text-align: center;
  height:15em;
  overflow: hidden;
  line-height: 2em;
}
h1 {
  margin-top: 30px;
  color: #533bff;
```

```
  text-shadow: 0 2px 0 #12cc20, 0 3px 0 #c9c9c9, 0 4px 0 #bb5453, 0 5px 0
  #b9b9b9, 0 6px 0 #aaa, 0 7px 2px rgba(0,0,0,0.1), 0 0 6px rgba(0,0,0,0.1),0
  2px 4px rgba(0,0,0,0.3),0 4px 5px rgba(0,0,0,0.2),0 6px 10px
  rgba(0,0,0,0.25);
}
.neonText h2 {
  color: #4789ff;
}
.slide-up-enter-active {
  transition: all .5s ease;
}
.slide-up-leave-active {
  transition: all .8s ease;
}
.slide-up-enter, .slide-up-leave-active {
  transform:scale(0,0);
}
</style>
```

14.4.2 播放组件

播放组件（Controls）通过按钮激活显示，在 song 组件中使用 $emit 触发 Controls 组件的自定义事件。

在谷歌浏览器控制台模拟，播放组件效果如图 14-5 所示。

图 14-5 播放组件效果

具体代码如下：

```
<template>
  <transition name="slide-fade">
  <div v-show="show" class="controls">
       <div class="audioplayer">
         <audio id="media" :src="song.mp3Url" style="width: 0px; height: 0px;
            visibility: hidden;" @timeupdate="handleTimeUpdate">
                 <source :src="song.mp3Url">
            </audio>
                  <div class="audioplayer-playpause" title="Play" @click="">
                    <a href="#">Play</a></div>
                       <div class="audioplayer-time audioplayer-time-
                          current">{{currentTimeStr}}</div>
            <div id="bar" class="audioplayer-bar">
               <div class="audioplayer-bar-loaded" :style="{width:
                  loadedPercent+'%'}" @click="changePos($event)"></div>
               <div class="audioplayer-bar-played" :style="{width:
                  playedPercent+'%'}" @click="changePos($event)"></div>
            </div>
                <div class="audioplayer-time audioplayer-time-
                  duration">{{durationStr}}</div>
            <div class="audioplayer-volume">
                 <div class="audioplayer-volume-button" title="Volume"><a
```

```
onclick="return false;">Volume</a></div>
                    <div class="audioplayer-volume-adjust">
                        <div id="volumeOuterBar" @click="changeVolume($event,1)">
                           <div id="volumeBar" :style="{height: volume+'px'}"></div>
                        </div>
                    </div>
                </div>
            </div>
       </div>
       </transition>
    </template>
    <script>
    import '../assets/css/audioplayer.css';
    import { mapState } from 'vuex';
    import { Toast } from 'mint-ui';
    import _ from 'lodash';
    var media;
    export default {
      name: 'controls',
      data () {
        return {
         show:false,
          lrcCurIndex:0,
          lrcLastIndex:0,
          currentTime:0,
          duration:1,
          loadedPercent:0,
          volume:100
        }
      },
      computed:{
        ...mapState({
          song:state=>state.song || {mp3Url:''},
          lyricArr:state=>state.lyricArr,
          lrcTimeArr:state=>state.lrcTimeArr
        }),
        currentTimeStr(){
          return this.convertToTime(this.currentTime);
        },
        durationStr(){
          return this.convertToTime(this.duration);
        },
        playedPercent(){
          return (this.currentTime/this.duration)*100;
        }
      },
      created(){
      },
      mounted(){
        console.log('song',this.song)
        media=document.getElementById("media");
        media.addEventListener("loadeddata",(e)=>{
          var interval = setInterval(()=>{
              if( media.buffered.length < 1 ) return true;
              this.loadedPercent= ( media.buffered.end( 0 ) / media.duration ) * 100;
              if( Math.floor( media.buffered.end( 0 ) ) >= Math.floor( media.duration
                ) ) clearInterval( interval );
          }, 100 );
        });
```

```
    media.addEventListener("ended",()=>{
      media.currentTime=0;
      this.$store.commit("changePlaying",false);
    });
  this.$root.$on('play',()=>{
      this.show=true;
      this.$store.commit("changePlaying",true);
      setTimeout(()=>{
        console.log("duration", media.duration)
        media.play();
      },10);

  });
  this.$root.$on('pause',()=>{
      media.pause();
      this.$store.commit("changePlaying",false);
    });
  },
  methods:{
    handleTimeUpdate(){
      if(this.duration!=0){
        this.duration=media.duration;
      }
      //console.log(this.durationStr)
      this.currentTime=media.currentTime;
      let curIndex=_.sortedIndex(this.lrcTimeArr, media.currentTime);
      if (this.lrcLastIndex == curIndex) {
          return;
      }
      this.lrcCurIndex = curIndex;
      this.color();
      this.disappear();
      this.lrcLastIndex = curIndex;
    },
    color() {
        for (var i = 0; i < this.lyricArr.length; i++) {
            this.lyricArr[i].selected = false;
        }
        if (this.lrcCurIndex > 0) {
            this.lyricArr[this.lrcCurIndex - 1].selected = true;
        }
    },
    disappear() {
        if (this.lrcCurIndex >= 2) {
            this.$store.commit("changeLrcMarginTop",this.lrcCurIndex*(-2)+4.25);
        }
    },
    changePos(e){
      console.log(e);
      let x=e.offsetX;
      console.log("x", x);
      console.log("totalWidth",document.getElementById("bar").offsetWidth);
      let percent=x/document.getElementById("bar").offsetWidth;
      console.log(percent)
      media.currentTime=this.duration*percent;
    },
    changeVolume(e,isOuter){
      let y=e.pageY;
```

```
        let totalWidth=document.getElementById("volumeOuterBar").offsetHeight;
        let offset=this.getOffset(document.getElementById("volumeOuterBar"));
        this.volume=Math.abs(y-(offset.top+totalWidth));
        media.volume=this.volume/100;
      },
      convertToTime(time){
        var min = Math.floor((time / 60) % 60);
        var sec = Math.floor(time % 60);
        var cTime;
        if(sec < 10) {
            sec = '0' + sec;
        }
        if(min<10){
          min='0'+min;
        }
        cTime = min + ':' + sec
        return cTime;
      },
      getOffset(Node, offset) {
          if (!offset) {
              offset = {};
              offset.top = 0;
              offset.left = 0;
          }
          if (Node == document.body) {//当该节点为body节点时，结束递归
              return offset;
          }
          offset.top += Node.offsetTop;
          offset.left += Node.offsetLeft;
          return this.getOffset(Node.parentNode, offset);//向上累加offset里的值
      }
    }
}
</script>
<!-- Add "scoped" attribute to limit CSS to this component only -->
<style scoped>
.controls{
}
</style>
```

14.4.3 歌曲信息组件

歌曲信息组件（Lyric.vue）中展示了歌曲的作曲、作词和时间。在谷歌浏览器控制台模拟，歌曲信息组件效果如图 14-6 所示。

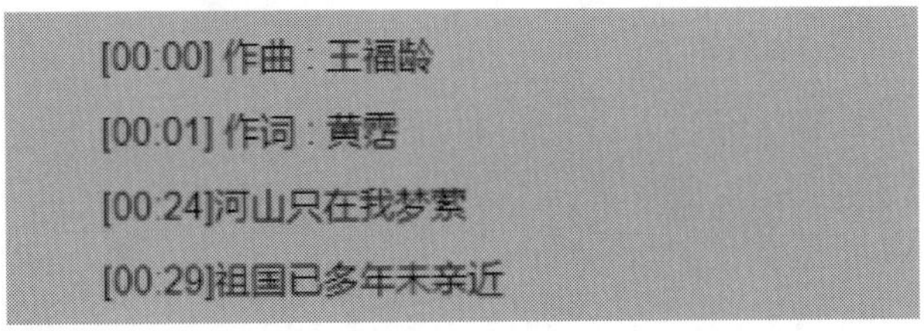

[00:00] 作曲 : 王福龄
[00:01] 作词 : 黄霑
[00:24]河山只在我梦萦
[00:29]祖国已多年未亲近

图 14-6　歌曲信息组件效果

具体的实现代码如下：

```
<template>
  <transition name="slide-fade">
  <div id="lyric" class="lyric">
    <ul :style="{marginTop:lrcMarginTop+'em'}">
      <li v-for="x of lrcArr" :key="x.time" :class="x.selected?selectedCol
        or:defaultColor" v-show="x.show">{{x.lrc}}</li>
    </ul>
  </div>
  </transition>
</template>
<script>
import { mapState } from 'vuex';
export default {
  name: 'lyric',
  props:['id'],
  data () {
    return {
      show:false,
      curIndex:0,
      defaultColor:'t-gra',
      selectedColor:'t-blu'
    }
  },
  computed:mapState({
    lrcArr: state=>state.lyricArr,
    lrcMarginTop:state=>state.lrcMarginTop
  }),
  mounted(){
    this.$store.dispatch("FETCH_LYRIC",this.id).then(()=>{
      this.show=true;
    });
  },
  methods:{
  }
}
</script>
<!-- 添加scoped属性以仅将CSS限制到此组件-->
<style scoped>
.lyric{
  height:8em;
  overflow-x:hidden;
  overflow-y: scroll;
}
.lyric li{
  list-style: none;
  line-height: 2em;
  transition:0.25s ease;
}
.t-gra{
  color: #0a0a0a;
}
.t-blu{
  color:#22c;
}
ul {
  transition: all 1s;
}
</style>
```

14.4.4 歌曲列表组件

歌曲列表组件（Songs.vue）列举出与搜索歌曲相关的列表。

在谷歌浏览器控制台模拟，Songs.vue 组件没有包含搜索框以及搜索历史，这里加上为了方便读者查看。项目启动后，输入“我的中国心”，如图 14-7 所示，然后单击“搜索”按钮，可进入歌曲列表页面，如图 14-8 所示。

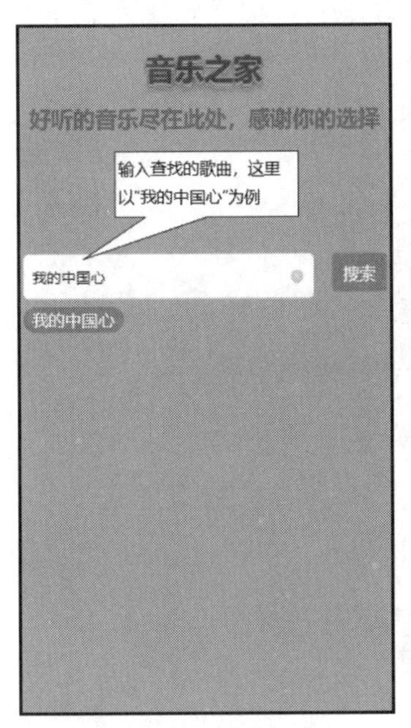

图 14-7　搜索歌曲效果

图 14-8　歌曲列表页面效果

具体代码如下：

```
<template>
  <div class="songs animation-style-1">
     <mt-cell v-if="show" class="item" :title="x.name" :label="x.artists[0].
        name" is-link :to="'/song/'+x.id" v-for="x of songs" :key="x.id">
        <span class="descp">Album {{x.album.name}}</span>
    </mt-cell>
  </div>
</template>
<script>
import { Indicator } from 'mint-ui';
import { mapState } from 'vuex';
export default {
  name: 'songs',
  data:function(){
    return {show:false};
  },
  computed:mapState(['songs']),
  created () {
    this.$store.commit("hideWelcome");
    this.$store.commit("changePlaying",false);
    this.fetchData()
  },
```

```
  watch: {
    '$route': 'fetchData'
  },
  methods:{
   fetchData(){
       console.log('fetchData')
     Indicator.open('加载中...');
        let word=this.$route.params.word || '汪峰';
      console.log(word)
       this.$store.dispatch("FETCH_SONG_LIST",{s:word}).then(()=>{
         Indicator.close();
         this.show=true;
       });
     }
  }
}
</script>
<!-- 添加scoped属性以仅将CSS限制到此组件-->
<style lang="less" rel="stylesheet/scss" scoped>
.descp{
  white-space:nowrap; overflow:hidden; text-overflow:ellipsis;
  max-width:10em;
}
.error {
  color: red;
}
.item {
    opacity: 0;
    animation-name: animationStyle1;
    animation-timing-function: ease-in-out;
    animation-fill-mode: forwards;
    cursor: pointer;
}
.shown-loop(@n, @i: 1) when (@i <=@n) {
    .item:nth-child(@{i}) {
        animation-duration: @i*200ms;
    }
    .shown-loop(@n, (@i + 1));
}
.shown-loop(14);
@keyframes animationStyle1 {
    0% {
        opacity: 0;
        transform: rotateY(-90deg) translate3d(0, 30px, 0);
    }
    100% {
        opacity: 1;
        transform: rotate(0deg) translate3d(0, 0, 0);
    }
}
</style>
```

14.4.5 歌曲详情组件

在歌曲列表组件中，选择一首歌，单击后可进入歌曲详情组件（Song.vue）。在歌曲的详情组件中，单击“播放”按钮，可显示播放组件内容，如图 14-9 所示。

图 14-9　播放组件内容

具体代码如下：

```
<template>
  <transition name="slide-fade">
  <div class="song f1 container-flex box-ver">
     <div  v-if="song" class="card f1">
       <div class="song-pic" :class="{rotating:playing}" :style="{backgroun
         dImage:'url('+song.album.picUrl+')'}">
          </div>
          <div class="song-info">
              <h4>{{song.name}}</h4>
              <div style="font-size:0.9em;">{{song.artists[0].name}}</div>
          </div>
          <div class="btn-controls container-flex">
              <div class="btn-con" @click="handlePlay">
                 <i class="fa fa-2x" :class="played" style="color:#090909;"></i>
              </div>

          </div>
          <Lyric :id="id"></Lyric>
          <controls class="my-controls"></controls>
     </div>
  </div>
  </transition>
</template>
<script>
import { Indicator } from 'mint-ui';
import Lyric from '../components/Lyric';
import Controls from '../components/Controls';
import { mapState } from 'vuex';
export default {
  name: 'song',
  data () {
    return {
```

```
        id:(this.$route.params.id || '307525'),
        show:false,
      }
    },
    computed:{
    ...mapState(['song','playing']),
    played(){
        return {'fa-play':!this.playing,'fa-stop':this.playing};
    }
    },
    created(){
    Indicator.open('加载中...');
    this.$store.commit("hideWelcome");
    this.$store.dispatch("FETCH_SONG",this.id).then(()=>{
        Indicator.close();
        this.show=true;
    });
    },
    methods:{
    handlePlay(){
        if(!this.playing){
            this.$root.$emit("play");

        }else{
            this.$root.$emit("pause");
        }
    }
    },
    components:{Lyric,Controls}
}
</script>
<!-- Add "scoped" attribute to limit CSS to this component only -->
<style scoped>
.song{
  padding-top: 0.2em;
  perspective: 1000px;
}
.song-pic{
  width:8em;
  height:8em;
  margin: 0 auto;
  background-repeat:no-repeat;
  background-size:contain;
  background-position: center;
  border-radius: 50%;
}
.song-info{
  text-align: center;
}
.card{
  transform-style: preserve-3d;
}
.btn-controls{
  text-align: center;
}
.btn-con{
  margin: 0.5em auto;
  background-color: #2c3e50;
  padding: 0.5em 0.8em;
```

```
  border-radius: 50%;
  opacity: 0.6;
}
.t-gra{
  color: #050505;
}
.t-blu{
  color:#22c;
}
.rotating{
  animation: rotate 30s linear 0s infinite normal both running;
}
.my-controls{
  width: 95%;
  position: absolute;
  left: 0;
  bottom: 0.1em;
  padding:0 .8em;
}
@keyframes rotate
{
from {transform:rotateY(0deg);}
to {transform:rotateY(360deg);}
}
@-webkit-keyframes rotate /*Safari and Chrome*/
{
from {transform:rotateY(0deg);}
to {transform:rotateY(360deg);}
}
</style>
```

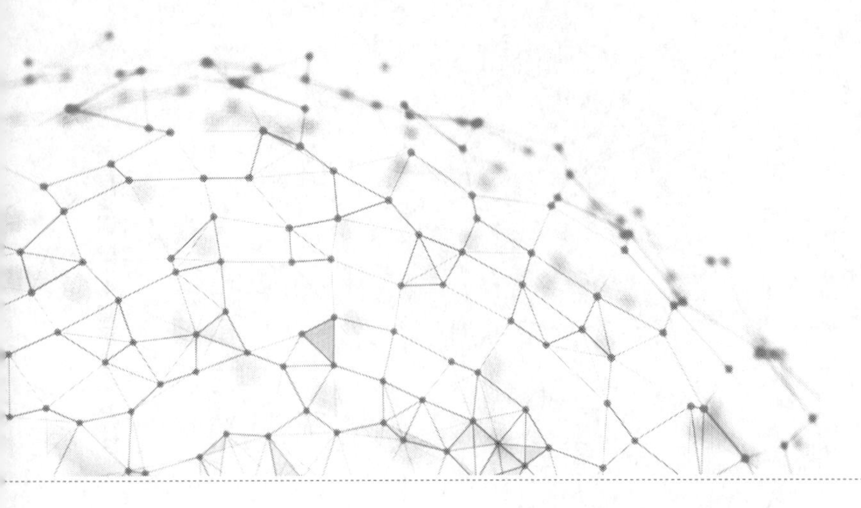

第15章

项目实训3——仿手机QQ页面

本章项目是一个移动端的仿手机 QQ 单页面应用，使用 Vue 框架设计开发。项目是使用 Vue-CLI 进行搭建，在项目中还使用了 Vue 的相关插件，例如：路由管理器（Vue-Router）、状态管理（Vuex）。项目使用的是模拟数据，使用 Axios 进行请求。

15.1 项目概述

项目是仿手机 QQ 单页面应用，实现了一些功能，例如对话功能、左滑删除、搜索好友等。

15.1.1 开发环境

首先需要安装 Node.js 和 NPM，一般情况下 Node.js 中已经集成了 NPM。然后安装 Vue 脚手架（Vue-Cli）以及创建项目，具体的安装步骤请参考“脚手架”这一章。

对于项目的调试，是在谷歌浏览器的控制台进行模拟。打开浏览器后，按下键盘上的 F12 键，然后单击“切换设备工具栏”，进入移动端的调试界面，可以选择相应的设备进行调试，如图 15-1 所示。

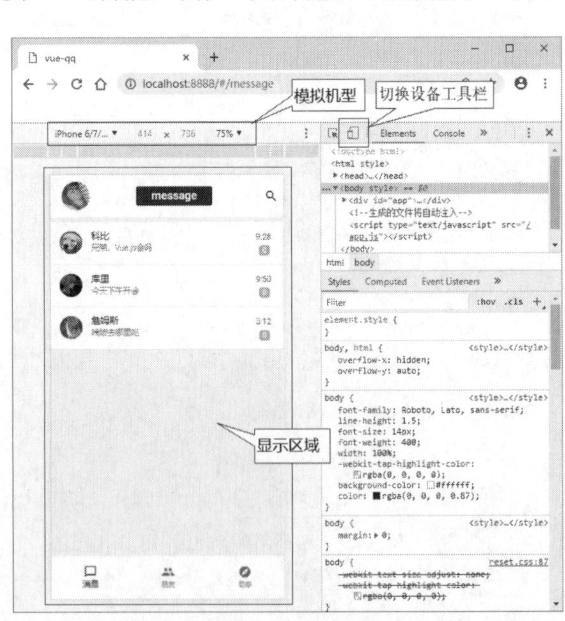

图 15-1　调试界面

15.1.2 技术概括

项目主要使用的技术说明如下。

- Axios：是一个基于 Promise 的 HTTP 客户端，专门为浏览器和 Node.js 服务。Vue 2.0 官方推荐使用 Axios 来代替原来的 Vue request。
- stylus：是 CSS 的预处理框架。stylus 给 CSS 添加了可编程的特性，在 stylus 中可以使用变量、函数、判断、循环一系列 CSS 没有的东西来编写样式文件，最后再把这个文件编译成 CSS 文件进行使用即可。
- webpack：webpack 是一个现代 JavaScript 应用程序的模块打包器 (module bundler)。当 webpack 处理应用程序时，它会递归地构建一个依赖关系图 (dependency graph)，其中包含应用程序需要的每个模块，然后将所有这些模块打包成一个或多个包。
- Muse-UI：Muse UI 是一套 Material Design 风格开源组件库，旨在快速搭建页面。它基于 Vue 2.0 开发，并提供了自定义主题，充分满足可定制化的需求。

15.1.3 项目结构

项目结构如图 15-2 所示，其中 src 文件夹是项目的源码目录，如图 15-3 所示。

图 15-2 项目结构

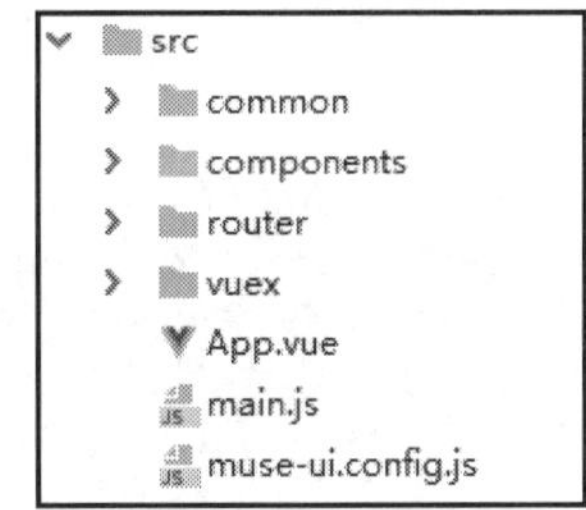

图 15-3 src 文件夹结构

项目结构中的主要文件说明如下。

- build：构建服务和 webpack 配置，转发聊天机器人及 Ajax 获取用户数据相关内容。
- config：项目不同环境的配置。
- dist：项目打包后的目录文件夹。
- index.html：项目入口页面。
- package.json：项目配置文件。
- mockdata.json：项目模拟的数据文件。

src 文件夹目录说明如下。

- common：公用 CSS 样式文件。
- components：项目组件文件夹。
- router：存放路由的文件夹。
- vuex：存放 Vuex 的相关文件。
  - muse-ui.config.js：Muse UI 单组件加载配置文件。
  - App.vue：模板入口文件。
  - main.js：程序入口文件。
  - static：CSS 文件、JavaScript 文件和图片资源文件夹。

15.2 入口文件

入口文件有 index.html、main.html 和 App.vue 三个文件。下面看一下具体的介绍。

15.2.1 项目入口页面

index.html 是项目默认的主渲染页面文件，主要是一些引用文件。具体代码如下：

```
<!DOCTYPE html>
<html>
  <head>
    <meta charset="utf-8">
    <title>vue-qq</title>
    <!--以下一些meta都是移动端常用的-->
    <meta name="screen-orientation" content="portrait">
    <meta name="full-screen" content="yes">
    <meta name="browsermode" content="application">
    <meta name="x5-orientation" content="portrait">
    <meta name="x5-fullscreen" content="true">
    <meta name="x5-page-mode" content="app">
      <meta name="viewport" content="width=device-width,initial-scale=1.0,
        minimum-scale=1.0, maximum-scale=1.0, user-scalable=no,minimal-ui">
    <link rel="stylesheet" type="text/css" href="static/css/reset.css">
    <!--以下两个链接是框架字体相关的-->
     <link rel="stylesheet" href="https://fonts.googleapis.com/css?family=Robot
o:300,400,500,700,400italic">
      <link rel="stylesheet" href="https://fonts.googleapis.com/
icon?family=Material+Icons">
  </head>
  <body>
    <div id="app"></div>
    <!--生成的文件将自动注入-->
  </body>
</html>
```

15.2.2 程序入口文件

main.js 是程序入口文件，加载各种公共组件以及初始化 Vue 实例。具体代码如下：

```
import Vue from 'vue'
import axios from 'axios'
import App from './App'
import router from './router/index'
import store from './vuex/store'
import MuseUi from './muse-ui.config'
Vue.use(MuseUi)
//懒加载模块
import VueLazyload from 'vue-lazyload'
Vue.use(VueLazyload, {
  preLoad: 1.3,
  error: 'static/images/lazy.jpg',
  loading: 'static/images/lazy.jpg',
  attempt: 1,
  listenEvents: ['scroll']
})
//在vue原型中添加$http方法等于axios
Vue.prototype.$http = axios
//设置默认打开的页面
router.replace('message')
Vue.config.productionTip = false
/* eslint-disable no-new */
new Vue({
  el: '#app',
  template: '<App/>',
  //注入路由
  router,
  //注入vuex的store
  store,
  components: { App },
  //组件创建前，进行异步数据请求
  beforeCreate() {
    this.$store.dispatch('getAllData', this)
  }
});
```

15.2.3 组件入口文件

App.vue 是项目的根组件，所有的页面都是在 App.vue 下面切换的，可以理解为所有组件都是 App.vue 的子组件。可以把每个页面都出现的文件（例如顶部、底部）都放在 App.vue 里面，还可以放公用的 CSS 代码。

```
<template>
  <div id="app">
    <!--头部导航-->
    <div class="container-top">
      <!--弹出层控制按钮-左边框弹出-->
      <div class="drawer"
           @click="showSidebar(true)"></div>
      <top-nav></top-nav>
```

```
      </div>
      <!--内容-->
      <div class="container-content">
        <div class="patch"></div>
        <router-view></router-view>
        <div class="patch"></div>
      </div>
      <!--尾部tab-->
      <div class="container-bottom">
        <bottom-tab class="tab"></bottom-tab>
      </div>
      <!--主页左侧弹出层-->
      <!--此处加z-index，防止弹出时被遮罩层遮挡-->
      <my-sidebar style="z-index: 20181224;"></my-sidebar>
      <!--消息界面-->
      <my-dialog class="my-dialog" v-if="dialog"></my-dialog>
      <!--个人主页-->
      <my-personindex class="my-personindex" v-if="personindex"></my-personindex>
      <!--搜索-->
      <my-search v-if="search"></my-search>
    </div>
</template>
<script>
import { mapState } from 'vuex'
//加载需要用到的组件
import bottomTab from './components/bottomtab/bottom-tab'
import topNav from './components/topnav/top-nav'
import myDialog from './components/dialog/dialog'
import mySidebar from './components/sidebar/sidebar'
import myPersonindex from './components/personindex/personindex'
import mySearch from './components/search/search'
export default {
  name: 'app',
  components: {
    bottomTab,
    topNav,
    myDialog,
    mySidebar,
    myPersonindex,
    mySearch
  },
  //mapState是vuex的方法之一
  computed: mapState(['dialog', 'personindex', 'search']),
  methods: {
    //点击左侧打开侧边栏
    showSidebar(flag) {
      this.$store.commit('showSidebar', { flag })
    }
  }
}
</script>
<style lang="stylus">
@import './common/stylus/mixin.styl'
#app
  .color-b
    color:color-b
  position:relative
  min-height: 100vh
  width: 100%
```

```
  background:color-g
  .my-dialog
    position: absolute
  .my-personindex
    .mu-icon.material-icons
      color: color-b
  .container-top
    position: fixed
    z-index: 101
    top: 0
    left: 0
    width: 100%
    height: 10%
    .drawer
      position: fixed
      z-index: 1
      top: 0
      left: 0
      width: 6vw
      height: 100%
  .container-bottom
    position: fixed
    z-index:1
    bottom: 0
    left: 0
    width: 100%
    height: 10%
  .container-content
    width:100%
    .patch
      position: relative
      top: 0
      left: 0
      width: 100%
      height: 10.2vh
    .t-1
      color: color-b
    .mu-tab-text.has-icon
      margin-top: -4px
    .i-1
      font-size: 2.5em
      color: #64dd17
    .i-2
      font-size: 2.5em
      color: #f44336
    .i-3
      font-size: 2.5em
      color: #00bfa5
    .ii-1
      font-size: 2.5em
      color: #ffd600
    .ii-2
      font-size: 2.5em
      color: #ec407a
    .ii-3
      font-size: 2.5em
      color: #2962ff
</style>
```

15.3 状态管理

15.3.1 action.js

actions 里存放的是异步操作。由于 Vuex 中的 state 的变更只能由 mutations 进行操作，所以 actions 不直接进行数据操作，而是调用 mutations 方法。下面代码中出现的 that 都是 Vue 实例对象，因为把 Axios 绑定在了 Vue 原型上，Vuex 无法调用，所以这里需要传入 this。

```
const actions = {
  //异步获取基础数据
  //这里使用了es7的async函数，相当于封装了promis的generator
  getAllData: async ({ commit }, that) => {
    //声明变量用来存放之后获得的数据
    let self = {}
    let friends = {}
    await that.$http.get('/api/self')
      .then(({ data }) => {
        self = data.data
      })
    await that.$http.get('/api/friends')
      .then(({ data }) => {
        friends = data.data
      })
    commit('getData', {
      self, friends
    })
  },
  //聊天机器人
  sendValue: async ({ commit }, { _id, message, that }) => {
    //声明一个变量用来储存ajax获取的数据
    let robotData = ''
    //处理输入的内容，设置self为true，作为一个标记
    commit('changeList', { self: true, _id, message })
    //进行ajax请求，此处的that是从组件内传来的对象this
    await that.$http.get('/api/robotapi', {
      params: {
        message,
        id: _id
      }
    }).then(res => {
      //将获取到的数据赋值给先前设置的变量
      robotData = JSON.parse(res.data.data)
    })
    //判断获取到的数据类型，再进行对应操作
    if (robotData.code === 100000) {
      commit('changeList', { _id, message: robotData.text })
    } else if (robotData.code === 200000) {
      let data = robotData.text + robotData.url
      commit('changeList', { _id, message: data })
    } else if (robotData.code === 302000) {
      commit('changeList', { _id, message: '暂不支持此类对话' })
```

```
    } else {
      commit('changeList', { _id, message: '暂不支持此类对话' })
    }
  }
}
export default actions
```

15.3.2 getters.js

getter 属性就像计算属性一样，getter 的返回值会根据它的依赖被缓存起来，且只有当它的依赖值发生了改变才会被重新计算。可以认为 getter 属性是 store 的计算属性。

getters.js 文件定义了以下内容。

- 根据当前选中的朋友的 _id 来筛选出当前的 friend 的具体数据。
- 对当前消息队列中的消息进行加工，添加对应的好友资料。
- 由 _id 筛选出对应的好友。

```
//类似计算数据，根据state的数据，筛选或者暴露一格新数据
const getters = {
  //根据当前选中的朋友的_id来筛选出当前的friend的具体数据
  friend: (state) => {
    return state.data.friends.filter(x => x._id === state.activeId)[0]
  },
  //对当前消息队列中的消息进行加工，添加对应的好友资料
  nowMessageList: (state) => {
    let list = state.messageList
    list.forEach(x => {
      //由_id筛选出对应的好友
      let friend = state.data.friends.filter(i => i._id === x._id)[0]
      x.friend = friend
    })
    return list
  }
}
export default getters
```

15.3.3 mutations.js

更改 store 中 state 状态的唯一方法就是提交 mutation。每个 mutation 都有一个字符串类型的事件和一个回调函数，需要改变 state 的值就要在回调函数中改变，要执行这个回调函数，需要执行一个相应的调用方法 store.commit。

Vuex 中 state 的数据只能被 mutations 方法所更改，本项目更改的内容都存放在 mutations.js 文件中。

```
const mutations = {
  //对话
  showDialog: (state) => {
    //判断当前动作是否在打开对话，如果是在打开对话，那么判断当前进行对话的好友
    //是否存在消息队列
```

```
      //如果当前活跃的好友不存在消息队列（被删除的）那么就恢复此好友的消息队列，如果存在，
      //那么无动作
    if (!state.dialog) {
      //用空数组来判断也是true，所以后面加个[0]
      let message = state.messageList.filter(x => x._id === state.activeId)[0]
      if (!message) {
        let oldMessage = state.messageListFB.filter(x => x._id === state.activeId)[0]
        state.messageList.splice(oldMessage._id - 1, 0, oldMessage)
      }
    }
    state.dialog = !state.dialog
  },
  //侧边栏
  showSidebar: (state, { flag } = {}) => {
    state.sidebar.open = !state.sidebar.open
    state.sidebar.docked = !flag
  },
  //个人主页
  showPersonindex: (state) => {
    state.personindex = !state.personindex
  },
  //搜索
  showSearch: (state) => {
    state.search = !state.search
  },
  //ajax获取到用户数据
  getData: (state, data) => {
    // 将ajax获取到的值赋予state
    state.data = data
    // ajax状态更改为结束
    state.isAjax = true
  },
  //标题
  changTitle: (state, { title }) => {
    state.headerTitle = title
  },
  //获取当前获得关注的朋友的_id
  getActiveId: (state, { activeId }) => {
    state.activeId = activeId
  },
  //消息队列、聊天队列处理
  changeList: (state, obj) => {
    let now = new Date()
    let time ='${now.getHours()}:${now.getMinutes()}'
    //判断信息是自己的还是ai的，然后插入聊天队列中
    if (obj.self) {
      //信息是自己发送的
      state.messageList.forEach((item, index, arr) => {
        if (item._id === obj._id) {
          obj._id = 0
          item.list.push({ ...obj, time })
        }
      })
    } else {
      //信息是ai发送的
      state.messageList.forEach((item, index, arr) => {
        if (item._id === obj._id) {
          item.list.push({ ...obj, time })
        }
```

```
        })
      }
    },
    //删除消息
    removeMessage(state, { _id }) {
      state.messageList.forEach((item, index, arr) => {
        //判断信息列表中id与正在删除的信息id是否相同，如果相同，就删除信息
        if (item._id === _id) {
          arr.splice(index, 1)
        }
      })
    }
  }
export default mutations
```

15.3.4 store.js

每一个 Vuex 应用的核心就是 store（仓库），它包含应用中大部分的状态 (state)。把上面定义的 action.js、getters.js、mutations.js 文件引入到 store.js 文件中，最后导出一个新生成的 store 对象。

```
import Vue from 'vue'
import Vuex from 'vuex'
import mutations from './mutations'
import getters from './getters'
import actions from './actions'
//注册vuex
Vue.use(Vuex)
//初始化一些常用数据，根据Vue的理念，使用到的数据都必须先进行初始化设置
let state = {
  //对话框
  dialog: false,
  //侧边栏
  sidebar: {
    open: false,
    docked: true
  },
  //用户主页
  personindex: false,
  //搜索框
  search: false,
  //导航栏标题
  headerTitle: 'message',
  //初始化基础数据
  data: { self: {}, friends: [] },
  //ajax请求数据是否结束
  isAjax: false,
  //当前被选中或者在聊天中的friend的_id
  activeId: 0,
  //聊天队列，这里为每个朋友添加了一个聊天队列，偷懒写法，如果有需要可以改成动态添加，_id
  //是作为聊天队列的标记，list是聊天内容，list里的数据格式{_id:xx, message:xxx}，组件内
  //会根据_id来将对话插入到左边或右边，判断message是自己还是ai发出的
  messageList: [
    {
      _id: 1,
      list: [{ _id: 1, message: '兄弟，Vue.js会吗', time: '9:28' }]
```

```
    }, {
      _id: 2,
      list: [{ _id: 2, message: '今天下午开会', time: '9:50' }]
    }, {
      _id: 3,
      list: [{ _id: 3, message: '晚饭去哪里吃', time: '3:12' }]
    }
  ],
  //消息队列副本，由于没有数据库，所以采用这样折中的方法
  messageListFB: [
    {
      _id: 1,
      list: [{ _id: 1, message: '兄弟，Vue.js会吗', time: '9:28' }]
    }, {
      _id: 2,
      list: [{ _id: 2, message: '今天下午开会', time: '9:50' }]
    }, {
      _id: 3,
      list: [{ _id: 3, message: '请问你要来点兔子吗', time: '3:12' }]
    }
  ]
}
//导出一个新生成的store对象
export default new Vuex.Store({
  state,
  mutations,
  actions,
  getters
})
```

15.4 项目组件及路由

项目的所有组件都在 components 文件夹中定义，路由在 router 文件夹中定义。

15.4.1 配置路由

在 router/index.js 的代码中，定义了路由相关的内容，会渲染到 App.vue 下面的 <router-view> 中。

```
import Vue from 'vue'
import Router from 'vue-router'
//注册router组件
Vue.use(Router)
//导入组件
import message from '../components/message/message.vue'
import friends from '../components/friends/friends.vue'
import discover from '../components/discover/discover.vue'
let routes = [
  { path: '/message', name: 'message', component: message },
  { path: '/friends', name: 'friends', component: friends },
  { path: '/discover', name: 'discover', component: discover }
```

```
]
export default new Router({
  routes
})
```

15.4.2 顶部导航栏组件

顶部导航栏组件（top-nav.vue）包含三个部分：人物头像、页面提示信息和搜索按钮。单击人物头像将进入侧边栏界面，单击搜索按钮将进入搜索界面。

在谷歌浏览器控制台中模拟，顶部导航效果如图 15-4 所示。

图 15-4　顶部导航效果

具体的实现代码如下：

```
<template>
  <div class="top-wrap">
    <mu-appbar class="top-nav"
               :zDepth="0">
      <!--等待添加弹出层按钮-->
      <mu-avatar slot="left"
                 :src="avatar"
                 :size="30"
                 @click="showSidebar_x(true)"  style="width: 50px;height: 50px;"/>

      <div slot="default"
           class="title">
        <div class="title-item">{{headerTitle}}</div>
      </div>
      <mu-icon slot="right"
               value="search"
               color="#2e2c6b"
               @click="showSearch" />
    </mu-appbar>
  </div>
</template>
<script>
import { mapState, mapMutations } from 'vuex'
export default {
  name: 'topNav',
  computed: mapState({
    avatar: state => state.data.self.avatar,
    headerTitle: 'headerTitle'
  }),
  methods: {
    ...mapMutations(['showSidebar', 'showSearch']),
    showSidebar_x(flag) {
      this.showSidebar({ flag })
    }
  }
```

```
}
</script>
<style lang="stylus" scoped>
@import '../../common/stylus/mixin.styl'
.mu-appbar
    position: absolute
    top: 0
    left: 0
    width: 100%
    height: 100%
    background: color-w
    .mu-avatar
      margin-left:12px
    .title
      padding-right: 12px
      .title-item
        margin: 0 auto
        width: 48%
        height: 34px
        line-height: 30px
        text-align: center
        border: 1px solid color-b
        border-radius: 4px
        font-weight: 500
        background: color-b
        color: color-w
</style>
```

15.4.3 侧边栏导航组件

在顶部导航栏组件中，单击人物头像可进入侧边栏导航界面。侧边栏导航组件（sidebar.vue）由一个进入个人信息页面的人物头像和一个列表组成。

在谷歌浏览器控制台中模拟，侧边栏导航效果如图 15-5 所示。

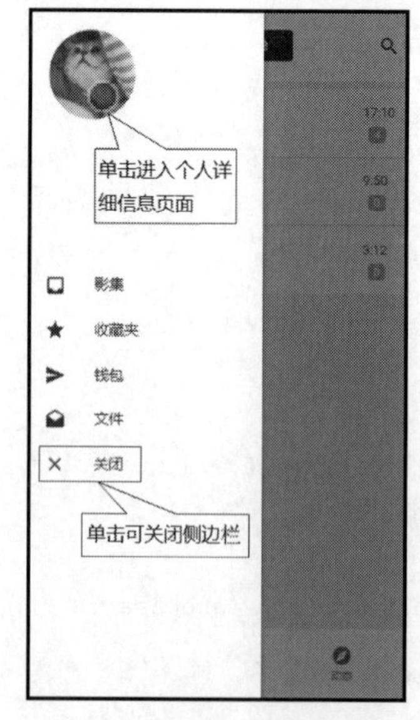

图 15-5　侧边栏导航效果

具体的实现代码如下：

```
<template>
  <mu-drawer :open="sidebar.open"
             :docked="sidebar.docked"
             @close="showSidebar()">
    <div class="content">
      <div class="top">
        <mu-avatar :src="self.avatar"
                   :size="96"
                   @click="showPersonindex_x" />
        <span class="name">{{self.name}}</span>
      </div>
      <div class="bottom">
        <span>{{self.explain}}</span>
      </div>
    </div>
    <mu-list-item title="影集">
      <mu-icon slot="left"
               value="inbox"
               color="#2e2c6b" />
    </mu-list-item>
    <mu-list-item title="收藏夹">
      <mu-icon slot="left"
               value="grade"
               color="#2e2c6b" />
    </mu-list-item>
    <mu-list-item title="钱包">
      <mu-icon slot="left"
               value="send"
               color="#2e2c6b" />
    </mu-list-item>
    <mu-list-item title="文件">
      <mu-icon slot="left"
               value="drafts"
               color="#2e2c6b" />
    </mu-list-item>
    <mu-list-item title="关闭"
                  @click.native="showSidebar">
      <mu-icon slot="left"
               value="close"
               color="#2e2c6b" />
    </mu-list-item>
  </mu-drawer>
</template>
<script>
import { mapState, mapMutations } from 'vuex'
export default {
  name: 'sidebar',
  computed: mapState({
    sidebar: 'sidebar',
    self: state => state.data.self
  }),
  methods: {
    ...mapMutations(['showSidebar', 'showPersonindex']),
    showPersonindex_x() {
      this.showSidebar()
      this.showPersonindex()
    }
  }
```

```
}
</script>
<style lang="stylus" scoped>
@import '../../common/stylus/mixin.styl'
.mu-drawer
  color: color-b
  .content
    position: relative
    height: 30vh
    padding-top: 1px
    margin-bottom: 50px
    .top
      padding: 20px
      .name
        position: absolute
        display: inline-block
        top: 5vh/*偏移*/
        left: 56% /*偏移*/
        font-size: 1.8em
    .bottom
      padding: 20px
      padding-left: 40px
</style>
```

15.4.4 搜索组件

在搜索组件（search.vue）中完成了搜索的功能。在顶部导航组件中，单击“搜索”按钮，可进入搜索界面。在搜索框中输入好友的名字，可以查找出朋友。

在谷歌浏览器控制台模拟，搜索界面效果如图 15-6 所示，在搜索框中输入“詹姆斯”，下面将显示好友的列表，如图 15-7 所示。

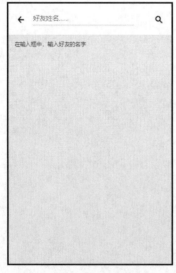

图 15-6　搜索界面效果

图 15-7　搜索结果

单击好友列表将跳转到好友的详细信息界面，如图 15-8 所示；在好友详细信息界面可以单击聊天信息图标，将进入和好友的聊天界面，如图 15-9 所示。

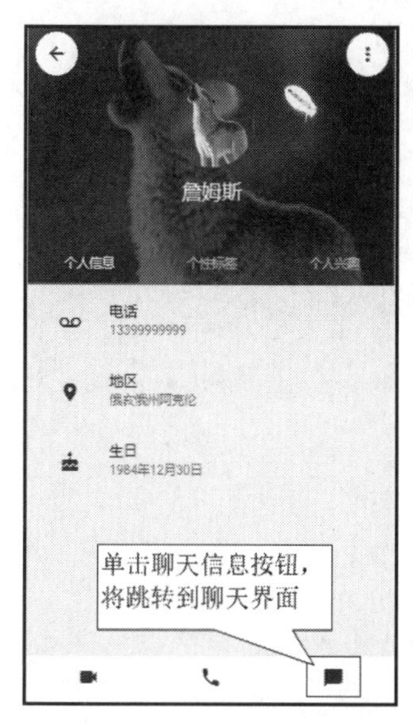

图 15-8　好友的详细信息界面

图 15-9　好友的聊天界面

具体实现代码如下：

```
<template>
  <div class="search">
    <mu-appbar :zDepth="0">
      <mu-icon-button icon="arrow_back"
                      slot="left"
                      @click="showSearch" />
      <mu-text-field class="appbar-search-field"
                     slot="default"
                     hintText="好友姓名......"
                     v-model="value"
                     @input="input" />
      <mu-icon-button icon="search"
                      slot="right" />
    </mu-appbar>
    <mu-list>
      <mu-sub-header>在输入框中，输入好友的名字</mu-sub-header>
      <div v-for="item of friend">
        <mu-list-item :title="item.name"
                      @click="showPersonindex_x(item._id)">
          <mu-avatar :src="item.avatar"
                     slot="leftAvatar" />
          <mu-icon value="chat_bubble"
                   slot="right" />
        </mu-list-item>
      </div>
    </mu-list>
  </div>
</template>
<script>
import { mapState, mapMutations } from 'vuex'
export default {
  name: 'search',
  data() {
    return {
```

```
        value: '',
        friend: []
      }
    },
    computed: {
      ...mapState({
        friends: state => state.data.friends
      })
    },
    methods: {
      ...mapMutations(['showSearch', 'showPersonindex', 'getActiveId']),
      //点击展示个人主页
      showPersonindex_x(id) {
        this.showSearch()
        this.getActiveId({ activeId: id })
        this.showPersonindex()
      },
      input(val) {
        //判断输入的值是否是数字
        if (val === '') {
          this.friend = []
        } else if (isNaN(val)) {
          //不是数字
          this.friend = this.friends.filter(x => {
            if (x.name.indexOf(val) !== -1) {
              return true
            } else {
              return false
            }
          })
        } else {
          //是数字
          this.friend = this.friends.filter(x => {
            if (x.phone.indexOf(val) !== -1) {
              return true
            } else {
              return false
            }
          })
        }
      }
    }
  }
</script>
<style lang="stylus" scoped>
@import '../../common/stylus/mixin.styl'
.search
  position: absolute
  z-index: 102
  top: 0
  left: 0
  width: 100%
  height: 100vh
  background: color-g
  .mu-appbar
    height: 10vh
    color: #000
    background: color-w
</style>
```

15.4.5 个人信息页面组件

个人信息页面组件（personindex.vue）由功能按钮、人物头像、tab栏和底部导航组成。功能按钮用来返回到消息界面，tab栏展示了个人信息、个性标签和个人兴趣等内容，底部导航并没有实现其功能。

在谷歌浏览器控制台中模拟，个人信息页面效果如图15-10所示，切换tab栏可以查看个性标签和个人兴趣的内容，分别如图15-11、图15-12所示。

图15-10 个人信息页面

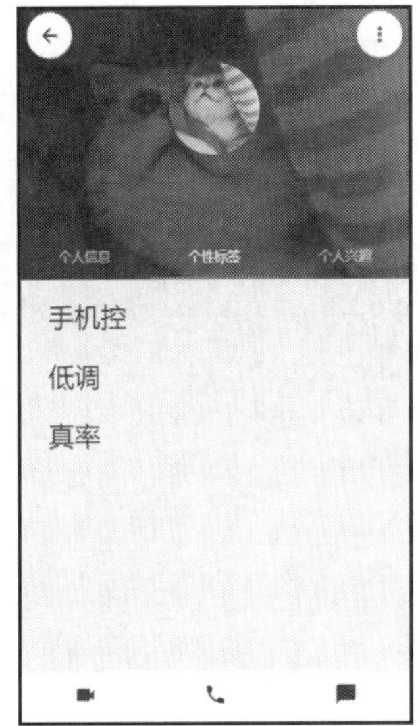

图15-11 个性标签页面

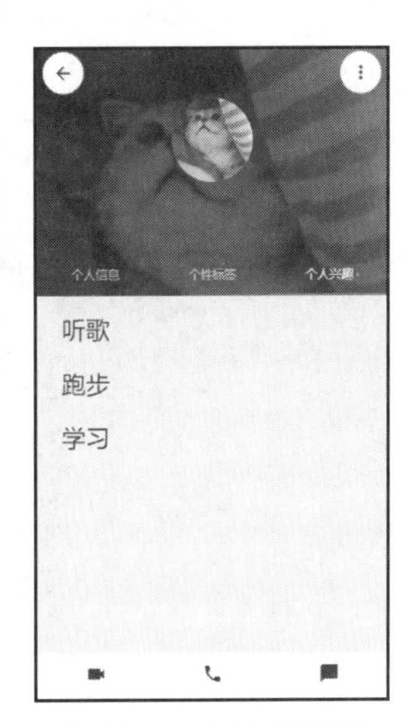

图15-12 个人兴趣页面

具体的实现代码如下：

```
<template>
  <div class="index">
    <div class="top"
         :style="{backgroundImage: 'url(${userData.avatar})'}">
      <mu-appbar :zDepth="0">
        <mu-icon-button icon="arrow_back"
                        slot="left"
                        @click="showPersonindex_x" style="background:white;"/>
        <div class="right-top"
             slot="right">
          <mu-icon-button icon="more_vert" style="background:white;"/>
        </div>
      </mu-appbar>
      <div class="c">
        <mu-avatar :src="userData.avatar"
                   :size="96" />
        <span class="name">{{userData.name}}</span>
      </div>
      <mu-tabs :value="activeTab"
               @change="handleTabChange">
        <mu-tab value="tab1"
                title="个人信息" />
        <mu-tab value="tab2"
                title="个性标签" />
```

```
        <mu-tab value="tab3"
                title="个人兴趣" />
      </mu-tabs>
    </div>
    <div class="content">
      <div class="item">
        <div v-if="activeTab === 'tab1'">
          <mu-list-item title="电话"
                        :describeText="userData.phone"
                        disabled>
            <mu-icon value="voicemail"
                     color="#2e2c6b"
                     slot="left"/>
          </mu-list-item>
        </div>
      </div>
      <div class="item">
        <div v-if="activeTab === 'tab1'">
          <mu-list-item title="地区"
                        :describeText="userData.address"
                        disabled>
            <mu-icon value="location_on"
                     color="#2e2c6b"
                     slot="left" />
          </mu-list-item>
        </div>
      </div>
      <div class="item">
        <div v-if="activeTab === 'tab1'">
          <mu-list-item title="生日"
                        :describeText="userData.birthday"
                        disabled>
            <mu-icon value="cake"
                     color="#2e2c6b"
                     slot="left" />
          </mu-list-item>
        </div>
      </div>
      <div v-if="activeTab === 'tab2'" style="padding-left:30px;">
        <h1>手机控</h1>
        <h1>低调</h1>
        <h1>真率</h1>
      </div>
      <div v-if="activeTab === 'tab3'" style="padding-left:30px;">
        <h1>听歌</h1>
        <h1>跑步</h1>
        <h1>学习</h1>
      </div>
    </div>
    <mu-tabs class="bottom" >
      <mu-tab value="tab1"
              icon="videocam" />
      <mu-tab value="tab2"
              icon="phone" color="#f00"/>
      <mu-tab value="tab3"
              icon="chat_bubble"
              @click="showDialog_x" />
    </mu-tabs>
  </div>
</template>
```

```
<script>
import { mapState, mapGetters, mapMutations } from 'vuex'
export default {
  name: 'personindex',
  data() {
    return {
      activeTab: 'tab1'
    }
  },
  computed: {
    ...mapState(['activeId', 'data']),
    ...mapGetters(['friend']),
    userData() {
      //判断是否有当前活跃的friend，没有的话就获取自己的数据，展示个人页面，有的话就展示
      //当前活跃朋友的页面
      if (this.activeId === 0) {
        return this.data.self
      } else {
        return this.friend
      }
    }
  },
  methods: {
    ...mapMutations(['getActiveId', 'showPersonindex', 'showDialog']),
    handleTabChange(val) {
      this.activeTab = val
    },
    showPersonindex_x() {
      this.getActiveId({ activeId: 0 })
      this.showPersonindex()
    },
    showDialog_x() {
      //判定打开的是否为自己的主页，如果是，则无法点击对话
      if (this.activeId !== 0) {
        this.showDialog()
        this.showPersonindex()
      }
    }
  }
}
</script>
<style lang="stylus" scoped>
@import '../../common/stylus/mixin.styl'
.icons
  color:#f00
.index
  position: absolute
  z-index: 102
  top: 0
  left: 0
  width: 100%
  height: 100vh
  background: color-g
  .top
    position: relative
    height: 38vh
    //background-image: url('./avatar.jpg')
    background-size: cover
    .c
      position: absolute
```

```
        z-index: 1
        width: 100%
        text-align: center
        .name
          display: block
          margin-top: 4px
          font-size: 1.6em
          color: #fff
      &:after
        position: absolute
        top: 0
        left: 0
        width: 100%
        height: 100%
        background: rgba(0,0,88,.5)
        content: ''
      .mu-appbar
        position: relative
        z-index:1
        background:rgba(0,0,0,0)
      .mu-tabs
        position: absolute
        bottom: 0
        left: 0
        z-index: 1
        background:rgba(0,0,0,0)
    .content
      .item
        margin-top: 6px
        margin-left: 20px
    .bottom
      position: absolute
      left: 0
      bottom: 0
      background: color-w
</style>
```

15.4.6 底部 tab 栏组件

底部 tab 栏组件（bottom-tab）用来切换消息界面、朋友界面和动态界面。在谷歌浏览器控制台中模拟，底部 tab 栏效果如图 15-13 所示。

图 15-13　底部 tab 栏组件

具体的实现代码如下：

```
<template>
  <mu-bottom-nav :value="bottomNav"
                 @change="handleChange"
                 class="bottom-tab">
    <mu-bottom-nav-item value="message"
```

```
                          title="消息"
                          icon="chat_bubble_outline"
                          :iconClass="[ isActive[0]&&'color-b' ]"
                          :titleClass="[ isActive[0]&&'color-b' ]" />
    <mu-bottom-nav-item value="friends"
                          title="朋友"
                          icon="people"
                          :iconClass="[ isActive[1]&&'color-b' ]"
                          :titleClass="[ isActive[1]&&'color-b' ]" />
    <mu-bottom-nav-item value="discover"
                          title="动态"
                          icon="explore"
                          :iconClass="[ isActive[2]&&'color-b' ]"
                          :titleClass="[ isActive[2]&&'color-b' ]" />
  </mu-bottom-nav>
</template>
<script>
export default {
  name: 'bottomTab',
  data() {
    return {
      bottomNav: 'message'
    }
  },
  computed: {
    isActive() {
      let arr = ['message', 'friends', 'discover']
      let x = []
      x[arr.indexOf(this.bottomNav)] = true
      return x
    }
  },
  methods: {
    //点击按钮
    handleChange(val) {
      this.bottomNav = val
      //路由跳转至当前点击的页面
      this.$router.push(val)
      //改变title
      this.$store.commit('changTitle', { title: val })
    }
  }
}
</script>
<style lang="stylus" scoped>
.bottom-tab
  position: absolute
  top: 0
  left: 0
  width: 100%
  height: 100%
</style>
```

提示

在后面的消息页面组件、朋友页面组件和动态页面组件效果展示中，为了读者更好地查看效果，把顶部导航组件和底部tab组件的效果也加进去了，但在消息页面组件、朋友页面组件和动态页面组件中并没有引入它们。

15.4.7 消息页面组件

在消息页面组件（message.vue）中，完成了左滑删除好友聊天记录的功能。

在谷歌浏览器控制台中模拟，效果如图 15-14 所示；将“库里”一列中向左滑动，显示“删除”按钮，效果如图 15-15 所示；点击“删除”按钮，好友聊天记录将被删除，效果如图 15-16 所示。

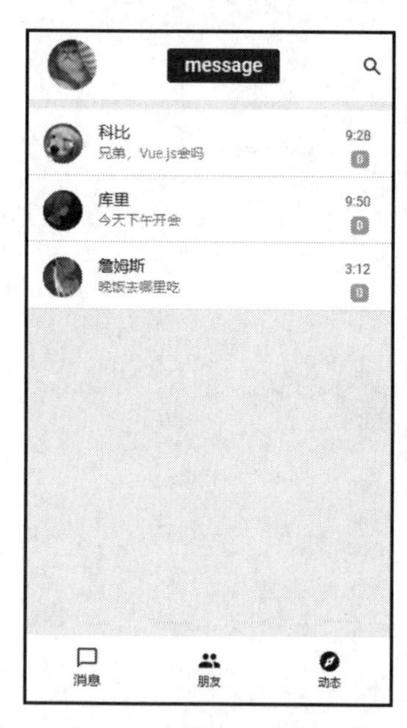

图 15-14 消息页面效果

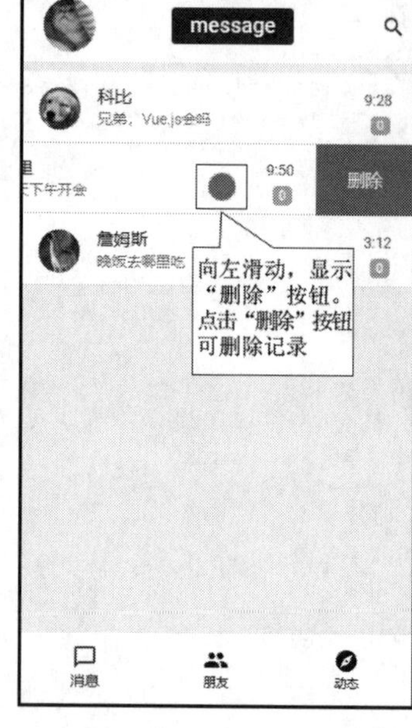

图 15-15 左滑好友效果

图 15-16 好友删除后效果

具体的实现代码如下：

```
<template>
  <!--判定ajax结束后，且有消息列表存在才开始渲染组件，防止报错-->
  <mu-list v-if="isAjax&&nowMessageList">
    <!--设置列表删除时动态效果-->
    <div v-for="(item, index) of nowMessageList"
         :class="[{swipeleft: isSwipe[index]},'wrap']"
         @click="getActiveId_x(item._id)"
         ref="child"
         :key="index">
      <mu-list-item :title="item.friend.name"
                    :describeLine="1"
                    :disableRipple="true"
                    class="list-item">
        <!--头像-->
        <mu-avatar :src="item.friend.avatar"
                   slot="leftAvatar" />
        <!--预览信息-->
        <span slot="describe">
          <span style="color: rgba(0, 0, 0, .5)">{{item.list[item.list.length-1].
           message}}</span>
        </span>
        <!--时间与待处理-->
```

```
        <div class="item-right"
             slot="right">
          <!--获取到当前聊天队列最后一条内容的time-->
          <span class="time">{{item.list[item.list.length-1].time}}</span>
          <!--数据条数-->
          <mu-badge :content="'${item.list.length-1}'" />
        </div>
      </mu-list-item>
      <!--分割线-->
      <!--阻止时间冒泡-->
      <div class="delete"
           @click.stop="removeM(item._id)">删除</div>
    </div>
  </mu-list>
</template>
<script>
import { mapState, mapGetters, mapMutations } from 'vuex'
export default {
  name: 'message',
  data() {
    return {
      isSwipe: [false, false, false]
    }
  },
  computed: {
    ...mapGetters(['nowMessageList']),
    //ajax是否已经结束
    ...mapState(['isAjax'])
  },
  methods: {
      ...mapMutations(['showDialog', 'getActiveId', 'zeroRemove',
        'removeMessage']),
    //获取点击的friend的_id
    getActiveId_x(id) {
      this.getActiveId({ activeId: id })
      this.showDialog()
    },
    //删除信息
    removeM(_id) {
      this.removeMessage({ _id })
    }
  },
  created() {
    setTimeout(() => {
      //判断是否存在信息列表
      if (this.$refs.child) {
        this.$refs.child.forEach((element, index) => {
          let x, y, X, Y, swipeX, swipeY
          //监听touchstart
          element.addEventListener('touchstart', e => {
            x = e.changedTouches[0].pageX
            y = e.changedTouches[0].pageY
            swipeX = true
            swipeY = true
            this.isSwipe = [false, false, false]
          })
          element.addEventListener('touchmove', e => {
            X = event.changedTouches[0].pageX
            Y = event.changedTouches[0].pageY
```

```
          if (swipeX && Math.abs(X - x) - Math.abs(Y - y) > 0) {
            //阻止默认事件
            e.stopPropagation()
            //右滑
            if (X - x > 10) {
              e.preventDefault()
              this.isSwipe.splice(index, 1, false)
            }
            if (x - X > 10) {
              e.preventDefault()
              this.isSwipe.splice(index, 1, true)
            }
            swipeY = false
          }
          if (swipeY && Math.abs(X - x) - Math.abs(Y - y) < 0) {
            swipeX = false
          }
        })
      })
    }
  }, 1000)
  }
}
</script>
<style lang="stylus">
@import '../../common/stylus/mixin.styl'
.mu-list
  overflow: hidden
  background: color-g
  //左滑删除
  .swipeleft
    transform:translateX(-20%)
  .wrap
    width: 125%
    overflow: hidden
    transition:all 0.3s linear
    border-b-1px(rgba(0,0,0,.1))
    .list-item
      float: left
      width:80%
      height: 10vh
      background: color-w
      transition:all 0.3s linear
    .delete
      float: right
      display: block
      height: 10vh
      line-height: 10vh
      width: 20%
      text-align: center
      font-size: 1.2em
      font-weight: 500
      color: color-w
      background: #ff1744
  .item-right
    position:relative
    .time
      display: inline-block
      position: absolute
```

```
      top: -10px
      left: -16px
    .mu-badge
      display: inline-block
      position: absolute
      top: 0
      left: -10px
      border-radius: 5px
</style>
```

15.4.8 聊天组件

在聊天组件（dialog.vue）中完成了对话功能。在项目中接入了图灵聊天机器人，可以与列表中的每个好友进行对话。这里把对话内容单独抽出来放到dialogue.vue组件中，内容如下。

dialogue.vue组件内容：

```
<template>
  <div>
    <div class="dialogue"
         v-for="item of messageList_x">
      <mu-list-item :disableRipple="true">
           <mu-avatar  :src="item._id===0?userData.self.avatar:userData.friend.
             avatar"
             :slot="item._id===0?'rightAvatar':'leftAvatar'" />
        <span :slot="item._id===0?'after':'title'">
            <span class="content" style="color: rgba(0, 0, 0, .9)">
              {{item.message}}</span>
        </span>
      </mu-list-item>
    </div>
  </div>
</template>
<script>
import { mapGetters } from 'vuex'
export default {
  name: 'dialogue',
  props: ['userData'],
  computed: {
    ...mapGetters(['nowMessageList']),
    //获取到当前的聊天记录
    messageList_x() {
      //筛选信息
        let message = this.nowMessageList.filter(x => x._id === this.userData.
          friend._id)[0]
      return message.list
    }
  },
  updated() {
    this.$emit('scrollC')
  }
}
</script>
<style lang="stylus" scoped>
@import '../../common/stylus/mixin.styl'
```

```
.dialogue
  margin-top: 10px
  font-size: 16px
  background: color-g
  .content
    display: inline-block
    padding: 1.5vh
    background: #fff
</style>
```

具体的聊天组件内容如下，在其中引入对话组件 dialogue.vue。

```
<template>
  <div class="dialog">
    <mu-appbar :title="userData.self.name"
              :zDepth="0">
      <mu-icon-button icon="arrow_back"
                      slot="left"
                      @click="showDialog_x" />
      <div class="right-top" slot="right">
        <mu-icon-button icon="videocam" />
        <mu-icon-button icon="call" />
        <mu-icon-button icon="person" @click="showPersonindex_x" />
      </div>
    </mu-appbar>
    <!--对话内容-->
    <div>
      <div class="patch-1"></div>
      <my-dialogue :userData="userData"
                  class="dialogue"
                  @scrollC="scrollC"></my-dialogue>
      <div class="patch-2"></div>
      <!--锚点-->
      <a name="1"
         href="#1"
         ref="end"
         style="height:0;color:rgba(0,0,0,0)">.</a>
    </div>
    <div class="footer"
         ref="footer">
      <div class="top">
        <mu-text-field hintText="输入文字"
                       v-model="value"
                       @focus="focus"
                       @blur="blur"
                       @keyup.enter.native="sendValue"/>
        <mu-icon-button icon="send" @click="sendValue"   />
      </div>
      <div class="bottom">
        <mu-icon-button icon="mic_none" />
        <mu-icon-button icon="photo_size_select_actual" />
        <mu-icon-button icon="tag_faces" />
        <mu-icon-button icon="switch_video" />
        <mu-icon-button icon="cloud_queue" />
        <mu-icon-button icon="photo_camera" />
        <mu-icon-button icon="folder_open" />
      </div>
    </div>
```

```
    </div>
  </template>
  <script>
  import { mapState, mapMutations } from 'vuex'
  import myDialogue from './dialogue'
  export default {
    name: 'dialog',
    components: {
      myDialogue
    },
    data() {
      return {
        value: '',
        timer: {}
      }
    },
    computed: {
      ...mapState({
        self: state => state.data.self,
        headerTitle: 'headerTitle'
      }),
      userData() {
        return {
          self: this.self,
          friend: this.$store.getters.friend
        }
      }
    },
    methods: {
      ...mapMutations(['showDialog', 'getActiveId', 'showPersonindex']),
      showDialog_x() {
        this.showDialog()
        this.getActiveId({ activeId: 0 })
        this.$router.push(this.headerTitle)
      },
      showPersonindex_x() {
        this.showDialog()
        this.showPersonindex()
        this.$router.push(this.headerTitle)
      },
      sendValue() {
        if (this.value.length) {
          this.$store.dispatch('sendValue', {
            _id: this.userData.friend._id,
            message: this.value,
            that: this
          })
        } else {
          console.log('不能为空')
        }
        this.value = ''
      },
      //监听子组件事件
      scrollC() {
        //取巧的方法，每次组件更新后模拟点击，破坏性地修改哈希值，但是简便(此处可以修改为
        //正常控制滚动条)
        this.$refs.end.click()
      },
      //输入框获得焦点时触发
```

```
      focus() {
        this.timer.T = setInterval(() => {
          //完美解决输入框被软键盘遮挡
          this.$refs.footer.scrollIntoView(false)
        }, 200)
      },
      blur() {
        //输入框失去焦点时清除定时器
        clearInterval(this.timer.T)
      }
    }
  }
</script>
<style lang="stylus" scoped>
@import '../../common/stylus/mixin.styl'
.dialog
  z-index: 999
  top: 0
  left: 0
  width: 100vw
  height: 100vh
  background: color-g
  .patch-1
    height: 60px
  .patch-2
    height: 90px
    background: color-g
  .mu-appbar
    position: fixed
    top: 0
    left: 0
    width: 100%
    background: color-w
    color: color-b
  .dialogue
    width: 100%
  .footer
    position: fixed
    bottom: 0
    left: 0
    width: 100%
    height: 90px
    text-align: center
    background: color-w
    .top
      display: flex
      justify-content: center
      padding: 0 10px
      .mu-icon-button
        display: inline-block
        margin-left: 18px
        vertical-align: top
    .bottom
      margin-top: -14px
      color:rgba(0,0,0,.3)
</style>
```

在谷歌浏览器控制台中模拟，输入相应的内容，会自动回复，如图 15-17 所示。点击右上角图标，可以查看好友的详细信息，如图 15-18 所示。

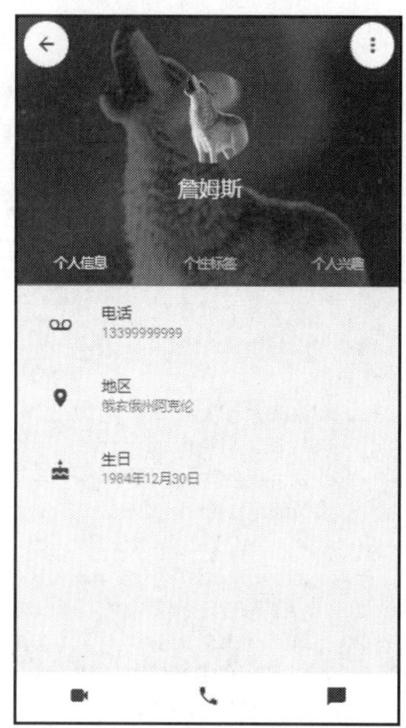

图 15-17　聊天页面效果　　图 15-18　好友详细信息

15.4.9　朋友页面组件

朋友页面组件（friends.vue）主要展示好友列表，还包括同步电话联系人、新朋友、群组、可能认识的人和我的动态等功能选项，本项目并没有实现其功能，有兴趣的读者可以试着完善。

在谷歌浏览器控制台中模拟，朋友页面效果如图 15-19 所示。点击好友“科比”可进入其详细信息界面，如图 15-20 所示。

图 15-19　朋友页面效果

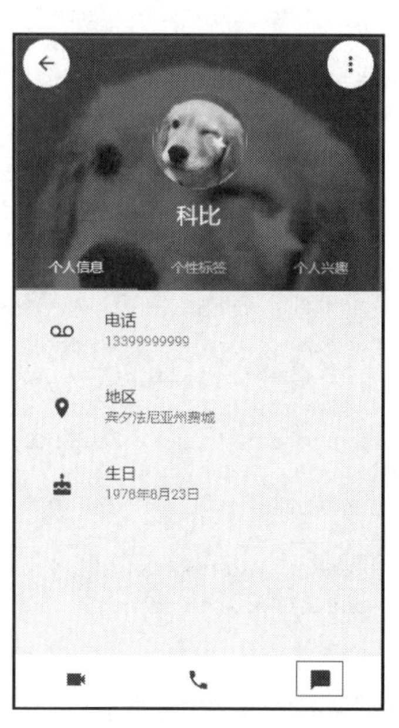

图 15-20　好友详细信息界面

具体实现代码如下：

```
<template>
  <div class="friend">
    <div class="title-1">
      <mu-icon value="assignment_ind" />
      <span class="text">同步电话联系</span>
      <mu-icon-button icon="chevron_right" />
    </div>
    <div class="tab">
      <mu-tabs>
        <mu-tab value="tab1"
                icon="person_add"
                title="新朋友"
                iconClass="i-1"
                titleClass="t-1"
                disabled/>
        <mu-tab value="tab2"
                icon="people"
                title="群组"
                iconClass="i-2"
                titleClass="t-1"
                disabled/>
        <mu-tab value="tab3"
                icon="person"
                title="可能认识的人"
                iconClass="i-3"
                titleClass="t-1"
                disabled/>
      </mu-tabs>
    </div>
    <div class="title-2">
      <span style="color: rgba(0, 0, 0, .8)">我的动态</span>
    </div>
    <div class="mac">
      <mu-list-item title="标签"
                    :describeLine="1"
                    :disableRipple="true">
        <mu-avatar icon="folder"
                   slot="leftAvatar"
                   color="#2e2c6b"
                   backgroundColor="#eee"
                   :size="40"
                   :iconSize="20" />
        <span slot="describe">
                    <span style="color: rgba(0, 0, 0, .5)">努力改变自己</span>
        </span>
        <div class="item-right"
             slot="right">
          <span class="time">2019.10.1</span>
        </div>
      </mu-list-item>
    </div>
    <div class="title-3">
      <span style="color: rgba(0, 0, 0, .8)">我的朋友</span>
    </div>
    <mu-list>
      <!--动态渲染-->
```

```
        <div v-for="item of friends" class="item">
          <mu-list-item :title="item.name"
                        @click="showPersonindex(item._id)"
                        :disableRipple="true">
            <mu-avatar :src="item.avatar"
                       slot="leftAvatar" />
            <mu-icon value="chat_bubble"
                     slot="right" />
          </mu-list-item>
          <mu-divider inset/>
        </div>
      </mu-list>
    </div>
</template>
<script>
export default {
  name: 'friend',
  computed: {
    //获取全部好友
    friends() {
      return this.$store.state.data.friends
    }
  },
  methods: {
    //点击展示个人主页
    showPersonindex(e) {
      this.$store.commit('getActiveId', { activeId: e })
      this.$store.commit('showPersonindex')
    }
  }
}
</script>
<style lang="stylus"  scoped>
@import '../../common/stylus/mixin.styl'
.friend
  .item
    background color-w
  .title-1
    position:relative
    height: 8vh
    line-height: 8vh
    text-align:center
    .mu-icon
      position: absolute;
      top:50%;
      left: 10%;
      transform:translate(-50%, -50%);
      color: color-b
    .mu-icon-button
      position: absolute;
      top:50%;
      left: 94%;
      transform:translate(-50%, -50%);
      color: color-b
    .text
      display: inline-block
      vertical-align: top
      font-size: 1.2em
  .tab
```

```
    height: 12vh
    overflow:hidden
    background: color-w
    .mu-tabs
      background:color-w
      color:color-b
  .title-2
    padding-left:4vw
    height: 6vh
    line-height: 6vh
  .mac
    position:relative
    height: 10vh
    background: color-w
  .title-3
    padding-left:4vw
    height: 4vh
    line-height: 5vh
</style>
```

15.4.10 动态页面组件

动态页面组件（discover.vue）由一个 tab 和一个列表组成。每一项的具体功能，这里没有实现，有兴趣的读者可以自行练习。

在谷歌浏览器控制台中模拟，动态页面效果如图 15-21 所示。

图 15-21　动态页面效果

具体的实现代码如下：

```
<template>
  <div class="friend">
    <div class="gap"></div>
    <div class="tab">
      <mu-tabs>
        <mu-tab value="tab1"
                icon="star"
                title="空间"
                iconClass="ii-1"
                titleClass="t-1"
                disabled/>
        <mu-tab value="tab2"
                icon="location_on"
                title="附近"
                iconClass="ii-2"
                titleClass="t-1"
                disabled/>
        <mu-tab value="tab3"
                icon="terrain"
                title="兴趣"
                iconClass="ii-3"
                titleClass="t-1"
                disabled/>
      </mu-tabs>
```

```
</div>

<div class="gap-1"></div>
<mu-list>
  <div class="item">
  <mu-list-item title="生活">
    <mu-icon slot="left"
             value="videocam"
             color="#64dd17" />
    <mu-icon value="keyboard_arrow_right"
             slot="right" />
  </mu-list-item>
  <mu-divider inset/>
  <mu-list-item title="视频">
    <mu-icon slot="left"
             value="video_library"
             color="#d84315" />
    <mu-icon value="keyboard_arrow_right"
             slot="right" />
  </mu-list-item>
  <mu-divider inset/>
  <mu-list-item title="音乐">
    <mu-icon slot="left"
             value="headset"
             color="#1976d2" />
    <mu-icon value="keyboard_arrow_right"
             slot="right" />
  </mu-list-item>
  </div>
</mu-list>

<mu-list>
  <div class="item">
  <mu-list-item title="图书">
    <mu-icon slot="left"
             value="book"
             color="#1e88e5" />
    <mu-icon value="keyboard_arrow_right"
             slot="right" />
  </mu-list-item>
  <mu-divider inset/>
  <mu-list-item title="新闻">
    <mu-icon slot="left"
             value="whatshot"
             color="#ef6c00" />
    <mu-icon value="keyboard_arrow_right"
             slot="right" />
  </mu-list-item>
  <mu-divider inset/>
  <mu-list-item title="游戏">
    <mu-icon slot="left"
             value="videogame_asset"
             color="#00e5ff" />
    <mu-icon value="keyboard_arrow_right"
             slot="right" />
  </mu-list-item>
  </div>
</mu-list>
<mu-list>
```

```
        <div class="item">
        <mu-list-item title="购物">
          <mu-icon slot="left"
                   value="shopping_cart"
                   color="#1e88e5" />
          <mu-icon value="keyboard_arrow_right"
                   slot="right" />
        </mu-list-item>
        </div>
      </mu-list>

  </div>
</template>
<script>
export default({
  name: 'discover'
})
</script>
<style lang="stylus"  scoped>
@import '../../common/stylus/mixin.styl'
.friend
  .item
    background: color-w
  .gap
    position:relative
    height: 2vh
  .gap-1
    height: 1vh
  .tab
    height: 12vh
    overflow:hidden
    background: colot-w
    .mu-tabs
      background:color-w
      color:color-b
</style>
```

15.5 模拟请求数据

本项目所使用的数据是模拟的数据，在 mockdata.json 文件中定义。具体内容如下：

```
{
    "self": {
        "_id": 0,
        "avatar": "/static/images/avatar.jpg",
        "birthday": "1963年2月17日",
        "name": "乔丹",
        "gender": "男",
        "email": "201990755@qq.com",
        "phone": "13399999999",
        "address": "美国纽约布鲁克林",
        "explain": "职业篮球运动员，司职得分后卫，绰号"飞人"",
```

```
            "about": "乔丹的职业生涯年年入选NBA全明星阵容（共14次）并3次当选NBA全明星
                MVP，10次入选NBA最佳阵容一阵，1985年入选NBA最佳阵容二阵，1988年荣膺NBA
                年度最佳防守球员，9次入选NBA最佳防守阵容一阵，3次荣膺NBA抢断王，2次夺得
                NBA全明星扣篮大赛冠军，1984年以及1992年夺得奥运会金牌"
        },
        "friend": [
            {    "_id": 1,
                "avatar": "/static/images/avatar1.jpg",
                "birthday": "1978年8月23日",
                "name": "科比",
                "gender": "男",
                "email": "201990755@qq.com",
                "phone": "13399999999",
                "address": "宾夕法尼亚州费城",
                "explain": "职业篮球运动员，司职得分后卫/小前锋（锋卫摇摆人），绰号"黑曼巴
                    "/"小飞侠"，是前NBA球员乔·布莱恩特的儿子。",
                "about": "科比是一名高产的得分手，他的职业生涯场均可以得到25分，还有5.2个篮
                    板、4.7次助攻和1.4次抢断， [34]  被认为是NBA最全面的球员之一"
            },
            {
                "_id": 2,
                "avatar": "/static/images/avatar2.jpg",
                "birthday": "1988年3月14日",
                "name": "库里",
                "gender": "男",
                "email": "201990755@qq.com",
                "phone": "13399999999",
                "address": "俄亥俄州阿克伦",
                "explain": "职业篮球运动员，司职控球后卫，效力于NBA金州勇士队",
                "about": "库里于2009年通过选秀进入NBA后一直效力于勇士队，新秀赛季入选最佳新
                    秀第一阵容；2014-15、2016-17、2017-18赛季三次获得NBA总冠军；两次荣膺常
                    规赛MVP，6次入选最佳阵容（3次一阵、2次二阵、1次三阵），6次入选全明星赛西
                    部首发阵容。"
            },
            {
                "_id": 3,
                "avatar": "/static/images/avatar3.jpg",
                "birthday": "1984年12月30日",
                "name": "詹姆斯",
                "gender": "男",
                "email": "201990755@qq.com",
                "phone": "13399999999",
                "address": "俄亥俄州阿克伦",
                "explain": "职业篮球运动员，司职小前锋，绰号"小皇帝"，效力于NBA洛杉矶湖
                    人队。",
                "about": "詹姆斯在2003年NBA选秀中于首轮第1顺位被克利夫兰骑士队选中，在2009
                    年与2010年蝉联NBA常规赛最有价值球员（MVP）。2010年，詹姆斯转会至迈阿密
                    热火队。2012年，詹姆斯得到NBA个人生涯的第3座常规赛MVP，第1个总冠军和总
                    决赛MVP，并代表美国男篮获得了伦敦奥运会金牌，追平了迈克尔·乔丹在1992年所
                    创的纪录。2013年，詹姆斯获得第4个常规赛MVP、第2个NBA总冠军和第2个总决赛
                    MVP，实现两连冠。2014年，詹姆斯回归骑士。2016年，詹姆斯带领骑士逆转战胜
                    卫冕冠军勇士获得队史首个总冠军和个人第3个总决赛MVP。"
            }
        ]
}
```

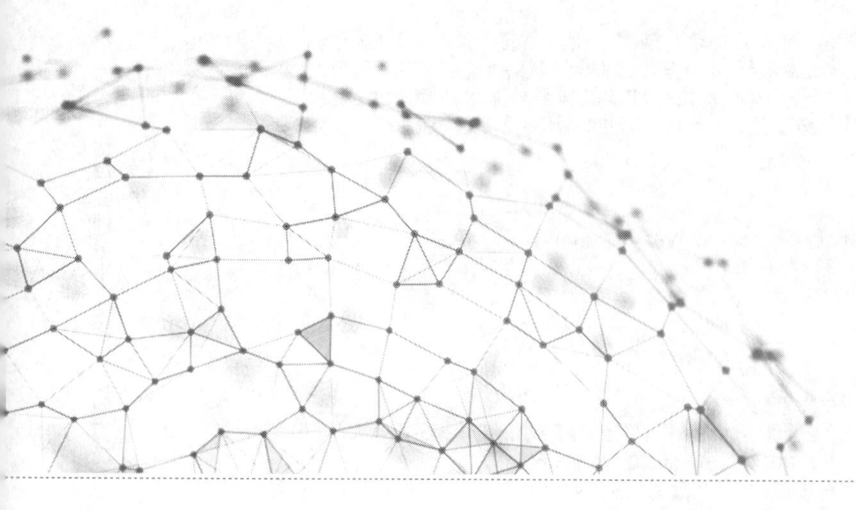

第16章 项目实训4——仿饿了么App

本章将介绍一款仿“饿了么”商家详情 App。它是基于 Vue 2.0 +Vue-Router + ES 6 +webpack 技术的一个 App，很适合读者进阶学习。

16.1 项目概述

项目是一款仿“饿了么”商家详情 App，主要实现了以下功能。

- 商品滚动，商品滚轮滚动。
- 商品联动。
- 加入购物车，移出购物车。
- 显示评论，评论筛选。
- 图片左右滑动。
- 商品详情，父子组件的通信。

16.1.1 开发环境

首先需要安装 Node.js 和 NPM，一般情况下 Node.js 中已经集成了 NPM。然后安装 Vue 脚手架（Vue-CLI）以及创建项目，具体的安装步骤请参考“脚手架”这一章。

项目的调试是在谷歌浏览器的控制台进行。打开浏览器后，按下键盘上的 F12 键，然后单击“切换设备工具栏”，进入移动端的调试界面，可以选择相应的设备进行调试，如图 16-1 所示。

图 16-1　项目调试效果

16.1.2　项目结构

项目结构如图 16-2 所示，其中 src 文件夹是项目的源码目录，如图 16-3 所示。

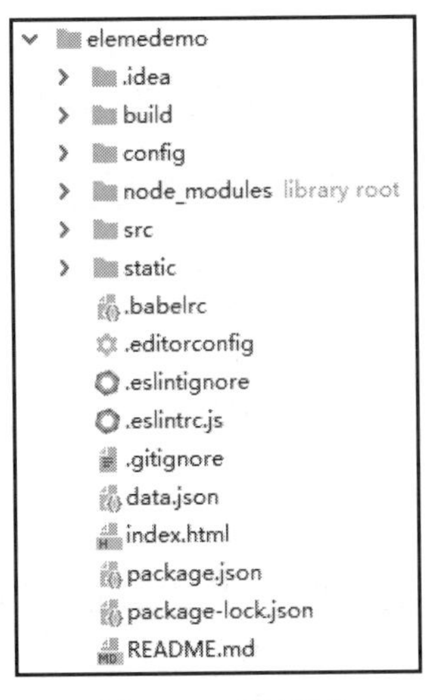

图 16-2　项目结构

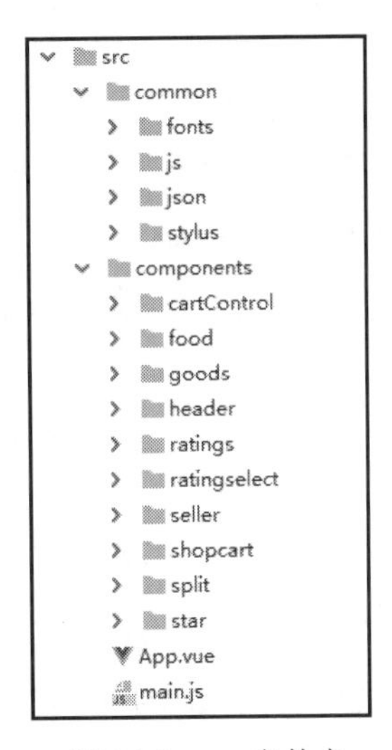

图 16-3　src 文件夹

项目结构中主要文件说明如下。

- build：构建服务和 webpack 配置。
- config：项目不同环境的配置。

- index.html：项目入口页面。
- package.json：项目配置文件。

src 文件夹目录说明如下。

- common：包括共用的字体图标、JavaScript 文件、模拟的 Json 数据和 CSS 样式。
- components：项目组件文件夹。
- App.Vue：模板入口文件。
- main.js：程序入口文件，其中包含了项目的路由设置。

16.2 入口文件

入口文件有 index.html、main.html 和 App.vue 三个文件。下面看一下具体的介绍。

16.2.1 项目入口页面

index.html 是项目默认的主渲染页面文件，主要是一些引用文件。具体代码如下：

```
<!DOCTYPE html>
<html>
  <head>
    <meta charset="utf-8">
    <title>eleme</title>
    <meta name="viewport" content="width=device-width,inital-scale=1.0,
    maximum-scale=1.0,user-scalable=no">
    <link rel="stylesheet" href="static/css/reset.css" type="text/css">
  </head>
  <body>
    <div id="app">
      <!-- route outlet -->
<!--路由匹配的组件将在这里呈现-->
      <router-view></router-view>
    </div>
  </body>
</html>
```

16.2.2 程序入口文件

main.js 是程序入口文件，加载各种公共组件以及初始化 Vue 实例，本项目路由的设置也在该文件中完成。具体代码如下：

```
import Vue from 'vue';
import VueRouter from 'vue-router';
import VueResource from 'vue-resource';
import App from './App';
import goods from './components/goods/goods.vue';
import ratings from './components/ratings/ratings.vue';
```

```
import seller from './components/seller/seller.vue';
import 'common/stylus/index.styl';
//安装 VueRouter这个插件
/* eslint-disable no-new */
Vue.use(VueRouter);
Vue.use(VueResource);
let routes = [
  {path: '/', name: 'index', component: App, children: [{path: '/goods',
     component: goods}, {path: '/ratings', component: ratings}, {path: '/
     seller', component: seller}]}
];
let router = new VueRouter({
  'linkActiveClass': 'active',
   routes //(缩写)相当于 routes: routes
});
let app = new Vue({
  router
}).$mount('#app');
  router.push('/goods');
export default app;
```

16.2.3 组件入口文件

App.vue 是项目的根组件，所有的页面都是在 App.vue 下面切换的，可以理解为所有组件都是 App.vue 的子组件。

```
<template>
  <div>
    <!-- 头部 -->
    <v-header :seller="seller"></v-header>
    <!-- 主体切换 -->
    <div class="tab border-1px">
      <div class="tab-item">
        <router-link v-bind:to="'/goods'">
          商品
        </router-link>
      </div>
      <div class="tab-item">
        <router-link to="/ratings">
          评论
        </router-link>
      </div>
      <div class="tab-item">
        <router-link to="/seller">
          商家
        </router-link>
      </div>
    </div>
    <!-- 头部 -->
    <keep-alive>
      <router-view :seller="seller"></router-view>
    </keep-alive>
  </div>
</template>
<script type="text/ecmascript-6">
  import header from './components/header/header.vue';
```

```
import {urlParse} from 'common/js/util';
import data from 'common/json/data.json';
export default {
  data() {
    return {
      seller: {},
      id: (() => {
        let queryParam = urlParse();
        console.log(queryParam);
        return queryParam.id;
      })()
    };
  },
  created() {
    this.seller = data.seller;
  },
  components: {
    'v-header': header
  }
};
</script>
<style lang="stylus" rel="stylesheet/stylus">
@import "common/stylus/mixin.styl";

.tab {
  display: flex;
  width: 100%;
  height: 40px;
  line-height: 40px;
border-1px(rgba(7, 17, 27, 0.1));
}
.tab .tab-item {
  flex: 1;
  text-align: center;
}
.tab .tab-item a {
  display: block;
  font-size: 14px;
  color: rgb(77, 85, 93);
}
.tab .tab-item .active {
  color: rgb(240, 20, 20);
}
</style>
```

16.3 项目组件

项目的所有组件都在 components 文件夹中定义，具体组件内容介绍如下。

16.3.1 头部组件

头部组件（header.vue）展示了商家的简单信息，如图 16-4 所示。当单击公告和“5

个”时，将显示详细的优惠信息和公告内容，如图 16-5 所示。

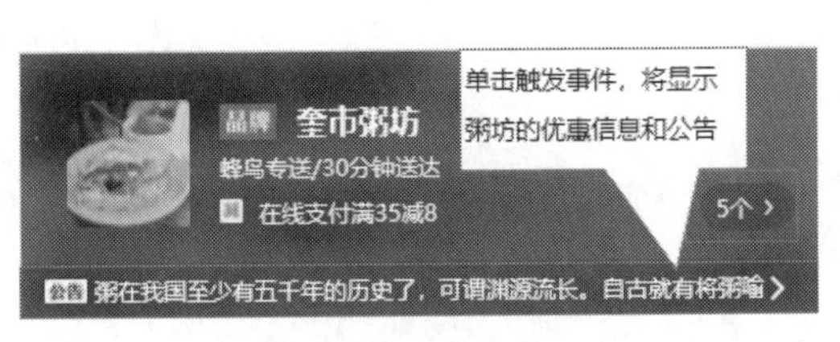

图 16-4　头部组件效果

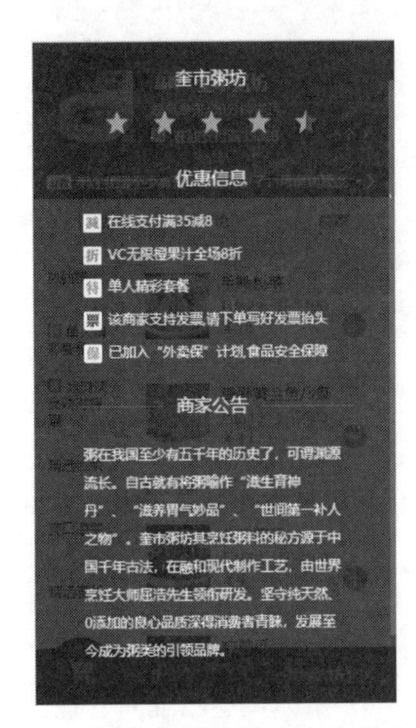

图 16-5　详细的优惠信息和公告内容

具体的实现代码如下：

```
<template>
  <div class="header">
    <div class="content-wrapper">
      <div class="avatar">
        <img width="64" height="64" :src="seller.avatar">
      </div>
      <div class="content">
        <div class="title">
          <span class="brand"></span>
          <span class="name">{{seller.name}}</span>
        </div>
        <div class="description">
          {{seller.description}}/{{seller.deliveryTime}}分钟送达
        </div>
        <div v-if="seller.supports" class="support">
          <span class="icon" :class="classMap[seller.supports[0].type]"></span>
          <span class="text">{{seller.supports[0].description}}</span>
        </div>
      </div>
      <div v-if="seller.supports" class="supports-count" @click="showDetail">
        <span class="count">{{seller.supports.length}}个</span>
        <i class="icon iconfont icon-zuoyoujiantou"></i>
      </div>
    </div>
    <div class="bulletin-wrapper" @click="showDetail">
      <span class="bulletin-title"></span><span class="bulletin-text">{{seller.
         bulletin}}</span>
      <i class="icon iconfont icon-zuoyoujiantou"></i>
    </div>
    <div class="background">
      <img :src="seller.avatar" alt="" class="" width="100%" height="100%">
    </div>
```

```
    <transition name="fade">
    <div v-show="detailShow" class="detail" @click="hideDetail"
     transition="fade">
     <div class="detail-wrapper clearFix">
       <div class="detail-main">
         <h1 class="name">{{seller.name}}</h1>
         <div class="star-wrapper">
           <star :size="48" :score="seller.score"></star>
         </div>
         <div class="title">
           <div class="line"></div>
           <div class="text">优惠信息</div>
           <div class="line"></div>
         </div>
         <ul v-if="seller.supports" class="supports">
           <li class="support-item" v-for="(item, index) in seller.supports">
             <span class="icon" :class="classMap[seller.supports[index].
               type]"></span>
             <span class="text">{{seller.supports[index].description}}</span>
           </li>
         </ul>
         <div class="title">
           <div class="line"></div>
           <div class="text">商家公告</div>
           <div class="line"></div>
         </div>
         <div class="bulletin">
           <p class="content">{{seller.bulletin}}</p>
         </div>
       </div>
     </div>
     <div class="detail-close" @click="hideDetail">
       <i class="iconfont icon-cha"></i>
     </div>
    </div>
    </transition>
  </div>
</template>
<script type="text/ecmascript-6">
  import star from '../star/star.vue';
  export default {
    props: {
      seller: {
        type: Object
      }
    },
    data() {
      return {
        detailShow: false
      };
    },
    methods: {
      showDetail() {
        this.detailShow = true;
      },
      hideDetail() {
        this.detailShow = false;
      }
    },
```

```
        created() {
            this.classMap = ['decrease', 'discount', 'special', 'invoice',
'guarantee'];
        },
        components: {
          star
        }
      };
    </script>
    <style lang="stylus" rel="stylesheet/stylus">
      @import "header.styl";
    </style>
```

16.3.2 商品数量控制组件

在商品数量控制组件（cartControl.vue）中，可以看到每列商品右下角有商品数量控制按钮，当单击按钮后，出现增加或减少购买数量的效果，如图16-6所示。

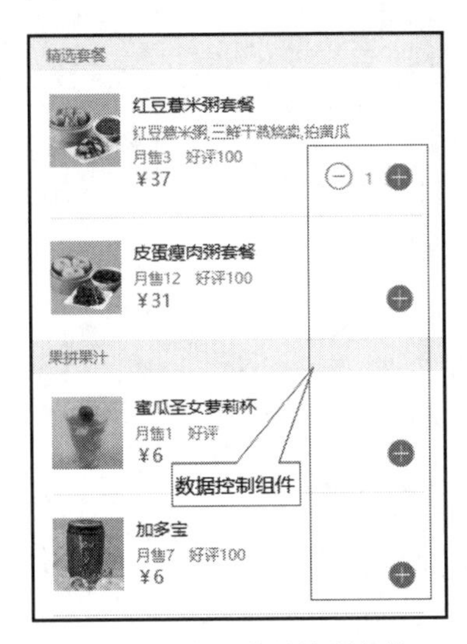

图16-6 商品数量控制组件

具体的实现代码如下：

```
<template>
  <div class="cartControl">
    <transition name="fade">
        <div class="cart-decrease" v-show="food.count>0" @click.stop.
          prevent="decreaseCart($event)">
          <transition name="inner">
          <span class="inner iconfont icon-jian"></span>
          </transition>
        </div>
    </transition>
     <span class="cart-count" v-show="food.count > 0 ">
       {{food.count}}
    </span>
    <span class="iconfont icon-jia cart-add" @click.stop.prevent="addCart($event)"></
       span>
```

```
    </div>
  </template>
  <script type="text/ecmascript-6">
    import Vue from 'vue';
    export default {
      props: {
        food: {
          type: Object
        }
      },
      methods: {
        addCart(event) {
          if (!event._constructed) {
            //去掉自带click事件的单击
            return;
          }
          if (!this.food.count) {
            Vue.set(this.food, 'count', 1);
          } else {
            this.food.count++;
          }
           //event.srcElement.outerHTML
             this.$emit('increment', event.target); //子组件通过 $emit触发父组件的方法increment
        },
        decreaseCart(event) {
          if (!event._constructed) {
            //去掉自带click事件的单击
            return;
          }
          this.food.count--;
        }
      }
    };
  </script>
  <style lang="stylus" rel="stylesheet/stylus">
    @import "cartControl.styl";
  </style>
```

16.3.3 购物车组件

在购物车组件（showcart.vue）中，在没有任何商品的情况下，是无法选择的，如图 16-7 所示。当选择商品后，购物车将被激活，如图 16-8 所示。这里实现了加入购物车和移除购物车的功能。

图 16-7 购物车默认状态

图 16-8 选择商品后效果

当单击购物车图标后，将显示我们选择的商品，如图 16-9 所示。在显示页面中可以增加或减少数量，也可以直接清空购物车。

当单击“去结算”按钮时将弹出购买商品花费的金额的提示对话框，如图 16-10 所示。

图 16-9 单击购物车显示商品

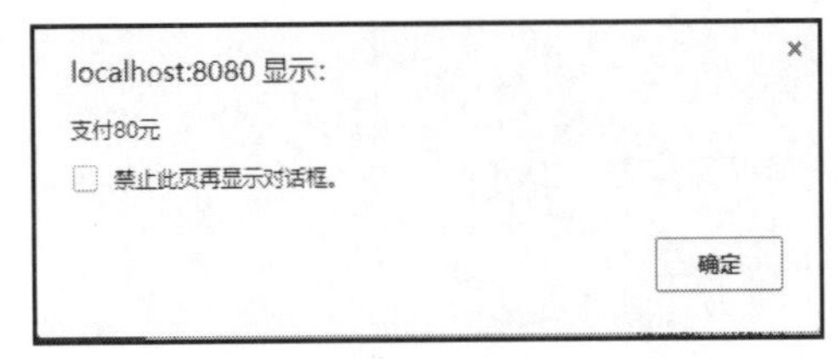

图 16-10 提示对话框

具体的实现代码如下：

```
<template>
  <div>
    <div class="shopCart">
      <div class="content" @click="toggleList($event)">
        <div class="content-left">
          <div class="logo-wrapper">
            <div class="logo" :class="{'highlight': totalCount > 0}">
              <i class="iconfont icon-gouwuche" :class="{'highlight':
                totalCount > 0}"></i>
            </div>
            <div class="num" v-show="totalCount > 0">{{totalCount}}</div>
          </div>
          <div class="price" :class="{'highlight': totalPrice >
            0}">¥{{totalPrice}}</div>
          <div class="desc">另需配送费¥{{deliveryPrice}}元</div>
        </div>
        <div class="content-right" @click.stop.prevent="pay">
          <div class="pay" :class="payClass">
            {{payDesc}}
          </div>
        </div>
      </div>
      <div class="ball-container">
        <div v-for="ball in balls">
          <transition name="drop" @before-enter="beforeEnter" @enter="enter" @
            after-enter="afterEnter">
            <div v-show="ball.show" class="ball">
              <div class="inner inner-hook">
              </div>
            </div>
          </transition>
        </div>
      </div>
      <transition name="fade">
        <div class="shopcart-list" v-show="listShow">
          <div class="list-header">
            <h1 class="title">购物车</h1>
            <span class="empty" @click="empty">清空</span>
          </div>
          <div class="list-content" ref="listContent">
            <ul>
              <li class="shopcart-food" v-for="food in selectFoods">
```

```
                <span class="name">{{food.name}}</span>
                <div class="price"><span>¥{{food.price * food.count}}</span></div>
                <div class="cartControl-wrapper">
                  <cartControl :food="food"></cartControl>
                </div>
              </li>
            </ul>
          </div>
        </div>
      </transition>
    </div>
    <transition name="fade">
      <div class="list-mask" v-show="listShow" @click="hideList()"></div>
    </transition>
  </div>
</template>
<script type="text/ecmascript-6">
  import cartControl from '../cartControl/cartControl.vue';
  import BScroll from 'better-scroll';
  export default {
    props: {
      selectFoods: {
        type: Array,
        default() {
          return [{price: 20, count: 2}];
        }
      },
      deliveryPrice: {
        type: Number,
        default: 0
      },
      minPrice: {
        type: Number,
        default: 0
      }
    },
    data () {
      return {
        balls: [{show: false}, {show: false}, {show: false}, {show: false},
          {show: false}],
        dropBalls: [],
        fold: true
      };
    },
    computed: {
      totalPrice() {
        let total = 0;
        this.selectFoods.forEach((food) => {
          total += food.price * food.count;
        });
        return total;
      },
      totalCount() {
        let count = 0;
        this.selectFoods.forEach((food) => {
          count += food.count;
        });
        return count;
      },
```

```
    payDesc() {
      if (this.totalPrice === 0) {
        return '¥${this.minPrice}元起送';
      } else if (this.totalPrice < this.minPrice) {
        let diff = this.minPrice - this.totalPrice;
        return '还差¥${diff}元起送';
      } else {
        return '去结算';
      }
    },
    payClass() {
      if (this.totalPrice < this.minPrice) {
        return 'not-enough';
      } else {
        return 'enough';
      }
    },
    listShow() {
      if (!this.totalCount) {
        this.fold = true;
        return false;
      }
      let show = !this.fold;
      if (show) {
        this.$nextTick(() => {
          if (!this.scroll) {
            this.scroll = new BScroll(this.$refs.listContent, {
              click: true
            });
          } else {
            this.scroll.refresh();
          }
        });
      }
      return show;
    }
  },
  methods: {
    toggleList() {
      if (!this.totalCount) {
        return;
      }
      this.fold = !this.fold;
    },
    empty() {
      this.selectFoods.forEach((food) => {
        food.count = 0;
      });
    },
    hideList() {
      this.fold = false;
    },
    pay() {
      if (this.totalPrice < this.minPrice) {
        return;
      }
      window.alert('支付' + this.totalPrice + '元');
    },
    drop(el) {
      for (let i = 0; i < this.balls.length; i++) {
```

```
            let ball = this.balls[i];
            if (!ball.show) {
              ball.show = true;
              ball.el = el;
              this.dropBalls.push(ball);
              return;
            }
          }
        },
        beforeEnter(el) {
          let count = this.balls.length;
          while (count--) {
            let ball = this.balls[count];
            if (ball.show) {
              let rect = ball.el.getBoundingClientRect();
              let x = rect.left - 32;
              let y = -(window.innerHeight - rect.top - 22);
              el.style.display = '';
              el.style.webkitTransform = 'translate3d(0,${y}px,0)';
              el.style.transform = 'translate3d(0,${y}px,0)';
              let inner = el.getElementsByClassName('inner-hook')[0];
              inner.style.webkitTransform = 'translate3d(${x}px,0,0)';
              inner.style.transform = 'translate3d(${x}px,0,0)';
            }
          }
        },
        enter(el) {
          this.$nextTick(() => {
            el.style.webkitTransform = 'translate3d(0,0,0)';
            el.style.transform = 'translate3d(0,0,0)';
            let inner = el.getElementsByClassName('inner-hook')[0];
            inner.style.webkitTransform = 'translate3d(0,0,0)';
            inner.style.transform = 'translate3d(0,0,0)';
          });
        },
        afterEnter(el) {
          let ball = this.dropBalls.shift();
          if (ball) {
            ball.show = false;
            el.style.display = 'none';
          }
        }
      },
      components: {
        cartControl
      }
    };
</script>
<style lang="stylus" rel="stylesheet/stylus">
  @import "shopCart.styl";
</style>
```

16.3.4 评论内容组件

评论内容组件（ratingselect.vue）有 4 个功能，分别为查看全部的评论内容、满意的内容、吐槽的内容和只查看有内容的评论，效果如图 16-11 所示。这里实现了显示评论和评论筛选的功能。

图 16-11　评论内容组件效果

具体的实现代码如下：

```
<template>
  <div class="ratingselect">
    <div class="rating-type border-1px">
      <span class="block positive" @click="select(2, $event)"
        :class="{'active': selectType === 2}">{{desc.all}}<span
        class="count">{{ratings.length}}</span> </span>
      <span class="block positive" @click="select(0, $event)"
            :class="{'active': selectType === 0}">{{desc.positive}}<span
        class="count">{{positives.length}}</span></span>
      <span class="block negative" @click="select(1, $event)"
            :class="{'active': selectType === 1}">{{desc.negative}}<span
        class="count">{{nagatives.length}}</span></span>
    </div>
    <div class="switch" @click="toggleContent( $event)"
      :class="{'on':onlyContent}">
      <i class="iconfont icon-gou"></i>
      <span class="text">只看有内容的评论</span>
    </div>
  </div>
</template>
<script type="text/ecmascript-6">
  const POSITIVE = 0;
  const NEGATIVE = 1;
  const ALL = 0;
  export default {
    props: {
      ratings: {
        type: Array,
        default() {
          return [];
        }
      },
      selectType: {
        type: Number,
```

```
        default: ALL
      },
      onlyContent: {
        type: Boolean,
        default: false
      },
      desc: {
        type: Object,
        default() {
          return {
            all: '全部',
            positive: '满意',
            negative: '吐槽'
          };
        }
      }
    },
    computed: {
      positives() {
        return this.ratings.filter((rating) => {
          return rating.rateType === POSITIVE;
        });
      },
      nagatives() {
        return this.ratings.filter((rating) => {
          return rating.rateType === NEGATIVE;
        });
      }
    },
    methods: {
      select (type, event) {
        if (!event._constructed) {
          return;
        }
        this.selectType = type;
        this.$emit('increment', 'selectType', type); //子组件通过 $emit触发父组件
        //的方法 increment，还可以传参
          this.$emit('increment' ,this.counter);
      },
      toggleContent (event) {
        if (!event._constructed) {
          return;
        }
        this.onlyContent = !this.onlyContent;
        this.$emit('increment', 'onlyContent', this.onlyContent);
      },
      needShow(type, text) {
        if (this.onlyContent && !text) {
          return false;
        }
        if (this.selectType === ALL) {
          return true;
        } else {
          return type === this.selectType;
        }
      }
    }
  };
</script>
```

```
<style lang="stylus" rel="stylesheet/stylus">
  @import "ratingselect.styl";
</style>
```

16.3.5 商品详情组件

在商品页面中，单击选择某件商品，将进入到商品的详情页面（food.vue）。在详情页面中，可以查看商品的大图展示效果以及买家对该商品的评论内容，如图16-12所示。还可以快捷地把商品加入到购物车中，效果如图16-13所示。

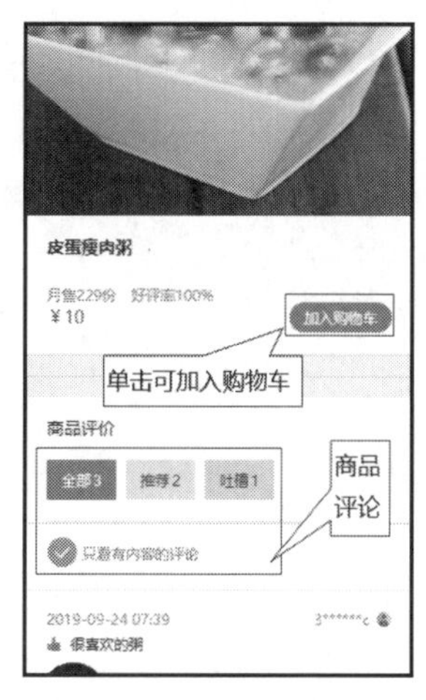

图16-12 商品详情组件效果

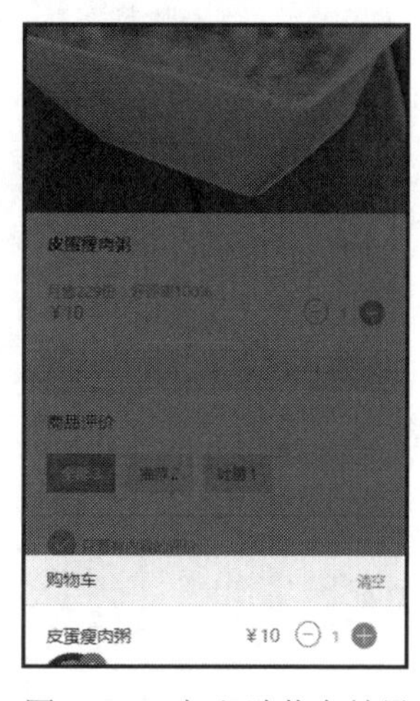

图16-13 加入购物车效果

具体的实现代码如下：

```
<template>
  <transition name="fade">
    <div v-show="showFlag" class="food">
      <div class="fond-content">
        <div class="image-header">
          <img :src="food.image" alt="">
          <div class="back" @click="hide">
            <i class="iconfont icon-weibiaoti6-copy"></i>
          </div>
        </div>
        <div class="content">
          <h1 class="title">{{food.name}}</h1>
          <div class="detail">
            <span class="sell-count">月售{{food.sellCount}}份</span>
            <span class="rating"> 好评率{{food.rating}}%</span>
          </div>
          <div class="price">
            <span class="now"> ¥{{food.price}}</span>
            <span class="old" v-show="food.oldPrice"> ¥{{food.oldPrice}}</span>
          </div>
          <div class="cartControl-wrapper">
            <cartControl :food="food"></cartControl>
          </div>
          <transition name="buy">
```

```
              <div class="buy" @click.stop.prevent="addFirst($event)"
                v-show="!food.count || food.count === 0">
                加入购物车
              </div>
            </transition>
          </div>
          <split></split>
          <div class="info" v-show="food.info">
            <h1 class="title">商品信息</h1>
            <p class="text">{{food.info}}</p>
          </div>
          <split></split>
          <div class="rating">
            <h1 class="title">商品评价</h1>
            <ratingselect @increment="incrementTotal" :select-type="selectType"
              :only-content="onlyContent" :desc="desc"
                    :ratings="food.ratings"></ratingselect>
            <div class="rating-wrapper">
              <ul v-show="food.ratings && food.ratings.length">
                <li v-show="needShow(rating.rateType, rating.text)"
                  class="rating-item border-1px"
                    v-for="rating in food.ratings">
                  <div class="user">
                    <span class="name">{{rating.username}}</span>
                    <img width="12" height="12" :src=rating.avatar alt=""
                      class="avatar">
                  </div>
                  <div class="time">{{rating.rateTime | formatDate}}</div>
                  <p class="text">
                    <i class="iconfont"
                      :class="{'icon-damuzhi':rating.rateType === 0,'icon-
                        down':rating.rateType === 1,}"></i>
                    {{rating.text}}
                  </p>
                </li>
              </ul>
              <div class="no-rating" v-show="!food.ratings || food.ratings.length
                === 0"></div>
            </div>
          </div>
        </div>
      </div>
    </transition>
</template>
<script type="text/ecmascript-6">
  import BScroll from 'better-scroll';
  import cartControl from '../cartControl/cartControl.vue';
  import split from '../split/split.vue';
  import ratingselect from '../ratingselect/ratingselect.vue';
  import Vue from 'vue';
  import {formatDate} from '../../common/js/date';
  const ALL = 2;
  export default {
    props: {
      food: {
        type: Object
      }
    },
    data () {
```

```
    return {
      showFlag: false,
      selectType: ALL,
      onlyContent: true,
      desc: {
        all: '全部',
        positive: '推荐',
        negative: '吐槽'
      }
    };
  },
  methods: {
    show() {
      this.showFlag = true;
      this.selectType = ALL;
      this.onlyContent = true;
      this.$nextTick(() => {
        if (!this.scroll) {
          this.scroll = new BScroll(this.$el, {
            click: true
          });
        } else {
          this.scroll.refresh();
        }
      });
    },
    incrementTotal(type, data) {
      this[type] = data;
      this.$nextTick(() => {
        this.scroll.refresh();
      });
    },
    hide() {
      this.showFlag = false;
    },
    addFirst(event) {
      if (!event._constructed) {
        return;
      }
      Vue.set(this.food, 'count', 1);
    },
    needShow(type, text) {
      if (this.onlyContent && !text) {
        return false;
      }
      if (this.selectType === ALL) {
        return true;
      } else {
        return type === this.selectType;
      }
    }
  },
  filters: {
    formatDate(time) {
      let date = new Date(time);
      return formatDate(date, 'yyyy-MM-dd hh:mm');
    }
  },
  components: {
```

```
        cartControl,
        ratingselect,
        split
      }
    };
</script>
<style lang="stylus" rel="stylesheet/stylus">
  @import "food.styl";
</style>
```

16.3.6 星级组件

星级组件（star.vue）是循环渲染字体图标，效果如图 16-14 所示。

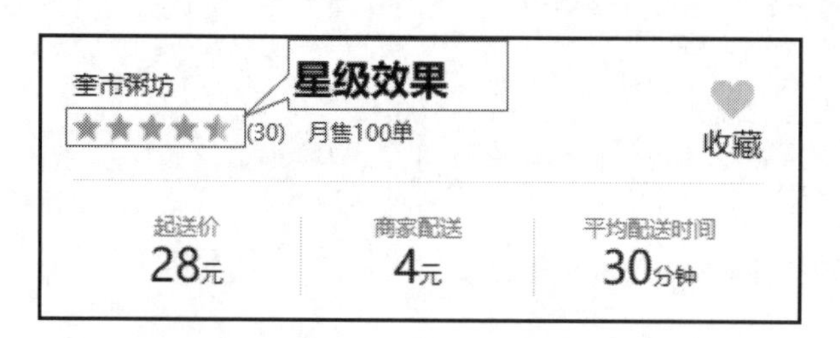

图 16-14　星级组件效果

具体的实现代码如下：

```
<template>
  <div class="star">
    <div class="star-item" :class="starType">
       <span v-for="itemClass in itemClasses" :class="itemClass" class="star-
item" ></span>
    </div>
  </div>
</template>
<script type="text/ecmascript-6">
  const LENGTH = 5;
  const CLS_ON = 'on';
  const CLS_HALF = 'half';
  const CLS_OFF = 'off';
  export default {
    props: {
      size: {
        type: Number
      },
      score: {
        type: Number
      }
    },
    computed: {
      starType() {
        return 'star-' + this.size;
      },
      itemClasses() {
        let result = [];
        let score = Math.floor(this.score * 2) / 2;
        let hasDecimal = score % 1 !== 0;
```

```
        let integer = Math.floor(score);
        for (let i = 0; i < integer; i++) {
          result.push(CLS_ON);
        }
        if (hasDecimal) {
          result.push(CLS_HALF);
        }
        while (result.length < LENGTH) {
          result.push(CLS_OFF);
        }
        return result;
      }
    }
  };
</script>
<style lang="stylus" rel="stylesheet/stylus">
  @import "star.styl";
</style>
```

16.3.7 商品组件

在商品组件（goods.vue）中，可以在左侧导航栏选择相应商品的类型，然后在列表中选择商品。本例使用数量控制组件来选择数量。这里为了方便读者查看效果，把头部组件的效果也加了进来，效果如图16-15所示。

图16-15 商品组件效果

具体的实现代码如下：

```
<template>
  <div class="good">
    <div class="menu-wrapper" ref="menuWrapper">
      <ul>
        <li v-for="(item, index) in goods" class="menu-item border-1px"
           :class="{'current':currentIndex === index}"
           @click="selectMenu(index, $event)">
          <span class="text">
            <span v-show="item.type>0" class=" icon" :class="classMap[item.
              type]"></span>{{item.name}}
          </span>
        </li>
      </ul>
    </div>
    <div class="foods-wrapper" ref="foodWrapper">
      <ul>
        <li v-for="item in goods" class="food-list food-list-hook">
          <h1 class="title">{{item.name}}</h1>
          <ul>
            <li v-for="food in item.foods" class="food-item" @
              click="selectFood(food, $event)">
              <div class="icon">
                <img :src="food.icon" alt="" width="57">
              </div>
              <div class="content">
                <h2 class="name">{{food.name}}</h2>
                <p class="desc">{{food.description}}</p>
```

```
                <div class="extra">
                  <span class="count">月售{{food.sellCount}}</span><span
                    class="count">好评{{food.rating}}</span>
                </div>
                <div class="price">
                  <span class="now">¥{{food.price}}</span><span class="old"
                   v-show="food.oldPrice">¥{{food.oldPrice}}</span>
                </div>
                <div class="cartControl-wrapper">
                  <cartControl :food="food" @increment="incrementTotal"></
                    cartControl>
                </div>
              </div>
            </li>
          </ul>
        </li>
      </ul>
    </div>
    <div>
      <shopCart :select-foods="selectFoods" :delivery-price="seller.
        deliveryPrice"
        :min-price="seller.minPrice" ref="shopCart"></shopCart>
      <food :food="selectedFood" ref="food"></food>
    </div>
  </div>
</template>
<script type="text/ecmascript-6">
  import BScroll from 'better-scroll';
  import shopCart from '../shopcart/shopCart.vue';
  import cartControl from '../cartControl/cartControl.vue';
  import food from '../food/food.vue';
  import data from 'common/json/data.json';
  //const ERR_OK = 0;
  export default {
    props: {
      seller: {
        type: Object
      }
    },
    data () {
      return {
        goods: [],
        listHeight: [],
        scrolly: 0,
        selectedFood: {}
      };
    },
    created() {
      this.goods = data.goods;
      this.$nextTick(() => {
        this._initScroll();
        this._calculateHeight();
      });
      this.classMap = ['decrease', 'discount', 'special', 'invoice',
        'guarantee'];
    },
    computed: {
      currentIndex() {
        for (let i = 0; i < this.listHeight.length; i++) {
```

```
        let height = this.listHeight[i];
        let height2 = this.listHeight[i + 1];
        if (!height2 || (this.scrolly >= height && this.scrolly < height2)) {
          return i;
        }
      }
      return 0;
    },
    selectFoods() {
      let foods = [];
      this.goods.forEach((good) => {
        good.foods.forEach((food) => {
          if (food.count) {
            foods.push(food);
          }
        });
      });
      return foods;
    }
  },
  methods: {
    _initScroll() {
      this.menuScroll = new BScroll(this.$refs.menuWrapper, {
        click: true
      });
      this.foodScroll = new BScroll(this.$refs.foodWrapper, {
        probeType: 3,
        click: true
      });
      this.foodScroll.on('scroll', (pos) => {
        this.scrolly = Math.abs(Math.round(pos.y));
      });
    },
    _calculateHeight() {
      let  foodList  =  this.$refs.foodWrapper.getElementsByClassName('food-
        list-hook');
      let height = 0;
      this.listHeight.push(height);
      for (let i = 0; i < foodList.length; i++) {
        let item = foodList[i];
        height += item.clientHeight;
        this.listHeight.push(height);
      }
    },
    selectMenu(index, event) {
      if (!event._constructed) {
        //去掉自带click事件的点击
        return;
      }
      let  foodList  =  this.$refs.foodWrapper.getElementsByClassName('food-
        list-hook');
      let el = foodList[index];
      this.foodScroll.scrollToElement(el, 300);
    },
    selectFood(food, event) {
      if (!event._constructed) {
        //去掉自带click事件的点击
        return;
      }
```

```
        this.selectedFood = food;
        this.$refs.food.show();
      },
      incrementTotal(target) {
        this.$refs.shopCart.drop(target);
      }
    },
    components: {
      shopCart,
      cartControl,
      food
    }
  };
</script>
<style lang="stylus" rel="stylesheet/stylus">
  @import "goods.styl";
</style>
```

16.3.8 评论组件

评论组件（ratings.vue）中包括两个方面的内容：商家综合评定和买家评论的内容，效果如图 16-16 所示。

图 16-16　评论组件效果

具体实现代码如下：

```
<template>
  <div class="ratings">
  <div>
    <div class="ratings-content">
      <div class="overview">
        <div class="overview-left">
          <h1 class="score">{{seller.score}}</h1>
          <div class="title">综合评分</div>
          <div class="rank">高于周边商家{{seller.rankRate}}%</div>
        </div>
        <div class="overview-right">
          <div class="score-wrapper">
            <span class="title">服务态度</span>
            <star :size="36" :score="seller.serviceScore"></star>
            <span class="score">{{seller.serviceScore}}</span>
          </div>
          <div class="score-wrapper">
            <span class="title">商品评分</span>
            <star :size="36" :score="seller.foodScore"></star>
            <span class="score">{{seller.foodScore}}</span>
          </div>
          <div class="delivery-wrapper">
            <span class="title">送达时间</span>
            <span class="delivery">{{seller.deliveryTime}}分钟</span>
          </div>
        </div>
      </div>
    </div>
    <split></split>
    <ratingselect   @increment="incrementTotal" :select-type="selectType"
```

```
        :only-content="onlyContent" :ratings="ratings"></ratingselect>
      <div class="rating-wrapper border-1px">
        <ul>
          <li v-for="rating in ratings" class="rating-item"
            v-show="needShow(rating.rateType, rating.text)">
            <div class="avatar">
              <img :src="rating.avatar" alt="" width="28" height="28">
            </div>
            <div class="content">
              <h1 class="name">{{rating.username}}</h1>
              <div class="star-wrapper">
                <star :size="24" :score="rating.score"></star>
                <span class="delivery" v-show="rating.deliveryTime">
                  {{rating.deliveryTime}}
                </span>
              </div>
              <p class="text">{{rating.text}}</p>
              <div class="recommend" v-show="rating.recommend &&rating.recommend.
                length">
                <i class="iconfont icon-damuzhi"></i>
                <span  class="item" v-for="item in rating.recommend" >{{item}}</span>
              </div>
              <div class="time">
                {{rating.rateTime | formatDate}}
              </div>
            </div>
          </li>
        </ul>
      </div>
    </div>
    </div>
</template>
<script type="text/ecmascript-6">
    import BScroll from 'better-scroll';
    import star from '../star/star.vue';
    import split from '../split/split.vue';
    import ratingselect from '../ratingselect/ratingselect.vue';
    import {formatDate} from '../../common/js/date';
    import data from 'common/json/data.json';
    const ALL = 2;
    export default {
      props: {
        seller: {
          type: Object
        }
      },
      data() {
        return {
          ratings: [],
          showFlag: false,
          selectType: ALL,
          onlyContent: true
        };
      },
      created() {
        this.ratings = data.ratings;
        this.$nextTick(() => {
          console.log(this.$el);
          this.scroll = new BScroll(this.$el, {click: true});
```

```
        });
      },
      methods: {
        incrementTotal(type, data) {
          this[type] = data;
          this.$nextTick(() => {
            this.scroll.refresh();
          });
        },
        needShow(type, text) {
          if (this.onlyContent && !text) {
            return false;
          }
          if (this.selectType === ALL) {
            return true;
          } else {
            return type === this.selectType;
          }
        }
      },
      filters: {
        formatDate(time) {
          let date = new Date(time);
          return formatDate(date, 'yyyy-MM-dd hh:mm');
        }
      },
      components: {
        star,
        split,
        ratingselect
      }
    };
</script>
<style lang="stylus" rel="stylesheet/stylus">
  @import "ratings.styl";
</style>
```

16.3.9 商家信息组件

在商家信息组件（seller.vue）中，设计了商家的星级和服务内容以及商家的优惠活动和公告内容，如图16-17所示。接下来还设计了商家实景以及商家的相关信息，如图16-18所示。在商家实景中实现了图片左右滑动的画廊功能。

具体的实现代码如下：

```
<template>
  <div class="seller">
    <div class="seller-content">
      <div class="overview">
        <h1 class="title">{{seller.name}}</h1>
        <div class="desc border-1px">
          <star :size="36" :score="seller.score"></star>
          <span class="text">({{seller.ratingCount}})</span>
          <span class="text">月售{{seller.sellCount}}单</span>
        </div>
        <ul class="remark">
```

```
<li class="block">
```

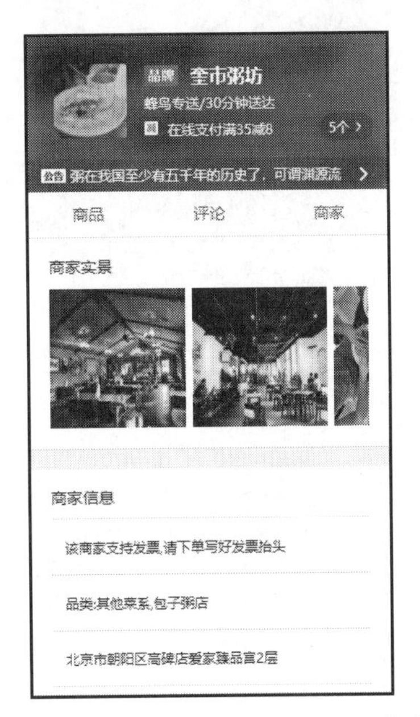

图 16-17　商家的星级和服务内容　图 16-18　商家实景以及商家的相关信息

```
      <h2>起送价</h2>
      <div class="content">
        <span class="stress">{{seller.minPrice}}</span>元
      </div>
    </li>
    <li class="block">
      <h2>商家配送</h2>
      <div class="content">
        <span class="stress">{{seller.deliveryPrice}}</span>元
      </div>
    </li>
    <li class="block">
      <h2>平均配送时间</h2>
      <div class="content">
        <span class="stress">{{seller.deliveryTime}}</span>分钟
      </div>
    </li>
  </ul>
  <div class="favorite" @click="toggleFavorite($event)">
    <i class="iconfont icon-aixin" :class="{'active':favorite}"></i>
    <span>{{favoriteText}}</span>
  </div>
</div>
<split></split>
<div class="bulletin">
  <h1 class="title">公告与活动</h1>
  <div class="content-wrapper border-1px">
    <p class="content">{{seller.bulletin}}</p>
  </div>
  <ul v-if="seller.supports" class="supports">
    <li class="support-item" v-for="(item, index) in seller.supports">
      <span class="icon" :class="classMap[seller.supports[index].
        type]"></span>
```

```
                <span class="text">{{seller.supports[index].description}}</span>
              </li>
            </ul>
          </div>
          <split></split>
          <div class="pics">
            <h1 class="title">商家实景</h1>
            <div class="pic-wrapper" ref="picWrapper">
              <ul class="pic-list" ref="picList">
                <li class="pic-item" v-for="pic in seller.pics">
                  <img :src="pic" width="120" height="120">
                </li>
              </ul>
            </div>
          </div>
          <split></split>
          <div class="info">
            <div class="title  border-1px">商家信息</div>
            <ul>
              <li class="info-item" v-for="info in seller.infos">{{info}}</li>
            </ul>
          </div>
        </div>
      </div>
    </template>
    <script type="text/ecmascript-6">
      import star from '../star/star.vue';
      import split from '../split/split.vue';
      import BScroll from 'better-scroll';
      import {savaToLocal, loadFromlLocal} from '../../common/js/store';
      export default {
        props: {
          seller: {
            type: Object
          }
        },
        components: {
          star,
          split
        },
        data() {
          return {
            favorite: (() => {
              return loadFromlLocal(this.seller.id, 'favorite', false);
            })()
          };
        },
        computed: {
          favoriteText() {
            return this.favorite ? '已收藏' : '收藏';
          }
        },
        created() {
          if (!this.picScroll) {
            if (this.seller.pics) {
              this.$nextTick(() => {
                let picWidth = 120;
                let margin = 6;
                let width = (picWidth + margin) * this.seller.pics.length - margin;
```

```
          this.$refs.picList.style.width = width + 'px';
          this.picScroll = new BScroll(this.$refs.picWrapper, {
            scrollX: true,
            eventPassthrough: 'vertical'
          });
        });
      }
    } else {
      this.picScroll.refresh();
    }
    if (!this.scroll) {
      this.$nextTick(() => {
        this.scroll = new BScroll(this.$el, {click: true});
      });
    } else {
      this.scroll.refresh();
    }
    this.classMap = ['decrease', 'discount', 'special', 'invoice',
      'guarantee'];
  },
  methods: {
    _initScroll() {
    },
    toggleFavorite(event) {
      if (!event._constructed) {
        return;
      }
      this.favorite = !this.favorite;
      savaToLocal(this.seller.id, 'favorite', this.favorite);
    }
  }
};
</script>
<style lang="stylus" rel="stylesheet/stylus">
  @import "seller.styl";
</style>
```